# HEALTHINFO 2024

The Ninth International Conference on Informatics and Assistive Technologies for Health-Care, Medical Support and Wellbeing

November 3rd - 7th, 2024

Nice, France

**HEALTHINFO 2024 Editors**

Katie Crowley, Lero - the Science Foundation Ireland Research Centre for Software and University of Limerick, Ireland

Ita Richardson, Lero - the Science Foundation Ireland Research Centre for Software and University of Limerick, Ireland

**Printed from e-media with permission by:**

Curran Associates, Inc.
57 Morehouse Lane
Red Hook, NY 12571

**Some format issues inherent in the e-media version may also appear in this print version.**



Printed with permission by Curran Associates, Inc. (2025)

International Academy, Research, and Industry Association (IARIA)
412 Derby Way
Wilmington, DE 19810

Phone: (408) 893-6407
Fax: (408) 527-6351

petre@iaria.org

**Additional copies of this publication are available from:**

Curran Associates, Inc.
57 Morehouse Lane
Red Hook, NY 12571 USA
Phone: 845-758-0400
Fax: 845-758-2633
Email: curran@proceedings.com
Web: www.proceedings.com

## Table of Contents

Real-world ADL Recognition with Deep Learning and Smartwatches: A Pilot Study 1
*Mustafa Elhadi Ahmed, Hongnian Yu, Michael Vassallo, and Pelagia Koufaki*

Challenges and Strategies of Inter-Disciplinary Research and Development: Lessons from a Telehealth Project 8
*Wei Zhou, Andy Li, Rashina Hoda, Chris Bain, Sam Georgy, Teresa O'Brien, Mahima Kalla, Olivia Metcalf, Wendy Chapman, and Peter Poon*

Hip Fracture Patient Pathways and Agent-based Modelling 11
*Alison N. O'Connor, Stephen E. Ryan, Gauri Vaidya, Paul Harford, and Meghana Kshirsagar*

User and Developer Perspectives of Technology Enabled Care 17
*Derek Ross, Ann-Marie Morrissey, Ita Richardson, Deepa Narasimha, Gobe Leinaeng Mphusu, Kevin Moore, and Katie Crowley*

Qawqaa: Aural Rehabilitation System for Children with Cochlear Implant Using Virtual Reality (VR) 23
*Leenah AlMashari, Sadeem AlSudais, and Meriam Kefi*

Fundamental Skills for Learning Strategies Based on Multiple Intelligences-Based on 192 Cases of Problematic Behavior in Vocational Training 26
*Norikatsu Fujita and Sho Aoki*

Mobile App Use among Chronic Pain Patients - A Gamified Approach 32
*Anabela Marques and Duarte Duque*

Transforming Healthcare: The Role of AI and Informatics in Modern Medical Support 37
*Amit Srivastava*

Social Engagement Embeddings of Parkinson's Disease through Autoencoders 43
*Ting Xiao, Poulomi Guha, and Ali Shah Tajuddin*

Does Complexity Pay Off? Applying Advanced Algorithms to Depression Detection on the GLOBEM Dataset 48
*Sebastian Cavada, Alvaro Berobide, and Yevheniia Kryklyvets*

AI-based Estimation of Lower Limb Joint Moments in Stance Phase Using a Single Wearable Inertial Sensor 53
*Kyoko Shibata and Kohei Watanabe*

# HEALTHINFO 2024

## Forward

The Ninth International Conference on Informatics and Assistive Technologies for Health-Care, Medical Support and Wellbeing (HEALTHINFO 2024), held on November 3 - 7, 2024 in Nice, France, tackles with particular aspects belonging to health informatics systems, health information, health informatics data, health informatics technologies, clinical practice and training, and wellbeing informatics in terms of existing and needed solutions.

The progress in society and technology regarding the application of systems approaches information and data processing principles, modeling and information technology, computation and communications solutions led to a substantial improvement of problems in assistive healthcare, public health, and the everyday wellbeing. While achievements are tangible, open issues related to global acceptance, costs models, personalized services, record privacy, and real-time medical actions for citizens' wellbeing are still under scrutiny.

We take here the opportunity to warmly thank all the members of the HEALTHINFO 2024 technical program committee as well as the numerous reviewers. The creation of such a broad and high quality conference program would not have been possible without their involvement. We also kindly thank all the authors that dedicated much of their time and efforts to contribute to the HEALTHINFO 2024. We truly believe that thanks to all these efforts, the final conference program consists of top quality contributions.

This event could also not have been a reality without the support of many individuals, organizations and sponsors. We also gratefully thank the members of the HEALTHINFO 2024 organizing committee for their help in handling the logistics and for their work that is making this professional meeting a success.

We hope the HEALTHINFO 2024 was a successful international forum for the exchange of ideas and results between academia and industry and to promote further progress in health informatics research. We also hope that Nice provided a pleasant environment during the conference and everyone saved some time for exploring this beautiful city

**HEALTHINFO 2024 General Chair**

Jaime Lloret Mauri, Polytechnic University of Valencia, Spain

**HEALTHINFO 2024 Steering Committee**

Shada Alsalamah, King Saud University, Saudi Arabia
Nelson P. Rocha, University of Aveiro, Portugal

**HEALTHINFO 2024 Publicity Chair**

Lorena Parra Boronat, Universitat Politecnica de Valencia, Spain
Sandra Viciano Tudela, Universitat Politecnica de Valencia, Spain
Jose Miguel Jimenez, Universitat Politecnica de Valencia, Spain
Francisco Javier Díaz Blasco, Universitat Politecnica de Valencia, Spain
Ali Ahmad, Universitat Politecnica de Valencia, Spain

# HEALTHINFO 2024

## Committee

### HEALTHINFO 2024 General Chair

Jaime Lloret Mauri, Polytechnic University of Valencia, Spain

### HEALTHINFO 2024 Steering Committee

Shada Alsalamah, King Saud University, Saudi Arabia
Nelson P. Rocha, University of Aveiro, Portugal

### HEALTHINFO 2024 Publicity Chair

Lorena Parra Boronat, Universitat Politecnica de Valencia, Spain
Sandra Viciano Tudela, Universitat Politecnica de Valencia, Spain
Jose Miguel Jimenez, Universitat Politecnica de Valencia, Spain
Francisco Javier Díaz Blasco, Universitat Politecnica de Valencia, Spain
Ali Ahmad, Universitat Politecnica de Valencia, Spain

### HEALTHINFO 2024 Technical Program Committee

Djafar Ould Abdeslam, University of Haute Alsace, France
Sherif Abdelwahed, Virginia Commonwealth University, USA
Somayeh Abedian, Ministry of Health and Medical Education, Tehran, Iran
Miriam Allalouf, Azrieli College of Engineering Jerusalem - JCE, Israel
Jens Allmer, Hochschule Ruhr West, University of Applied Sciences, Germany
João R. Almeida, University of Aveiro, Portugal / University of A Coruña, Spain
Shada Alsalamah, King Saud University, Saudi Arabia
Mustapha Aouache, Telecom Division - Centre de Développement des Technologies Avancées (CDTA), Algiers, Algeria
Khalfalla Awedat, Pacific LutheranUniversity, USA
Mana Azarm, University of Ottawa, Canada
Nabil Georges Badr, Higher Institute of Public Health – USJ, Beirut, Lebanon
Panagiotis D. Bamidis, Aristotle University of Thessaloniki, Greece
Hugo Barbosa, University Lusófona, Portugal
Fabio Baselice, University of Naples Parthenope, Italy
Arriel Benis, Holon Institute of Technology, Israel
Ahmed Bentajer, National School of Applied Sciences | Abdelmalek Essaad University, Tetouan, Morocco
Vilmos Bilicki, University of Szeged, Hungary
Amine Boufaied, ISITCom | University of Sousse, Tunisia
Guillaume Bouleux, University of Saint Etienne | INSA-Lyon, France
Klaus Brinker, Hamm-Lippstadt University of Applied Sciences, Germany
Tolga Çakmak, Hacettepe University, Turkey

Manuel Campos Martínez, University of Murcia, Spain
Armand Castillejo, STMicroelectronics, France
Rui Pedro Charters Lopes Rijo, Polytechnic of Leiria | INESCC | CINTESIS, Portugal
K.A.D. Chathurangika P. Kahandawaarachchi,Sri Lanka Institute of Information Technology, Sri Lanka
Ayan Chatterjee, University of Agder, Grimstad, Norway
Sudarshan S. Chawathe, University of Maine, USA
Bhargava Chinni, University of Rochester, USA
Giulia Cisotto, University of Padova, Italy / National Centre for Neurology and Psychiatry of Tokyo, Japan
Alberto Cliquet Jr., UNICAMP / USP, Brazil
Zhou Congcong, Zhejiang University, China
Andrea Corradini, KEA Copenhagen, Denmark
Katie Crowley, University of Limerick, Ireland
Subhashis Das, Dublin City University, Ireland
Giuseppe De Pietro, Institute for High Performance Computing and Networking (ICAR) - Italian National Research Council (CNR), Italy / Temple University's College of Science and Technology, Philadelphia, USA
Huseyin Demirci, University of Luxembourg, Luxembourg
Steven A. Demurjian, The University of Connecticut, USA
Jun-En Ding, Stevens Institute of Technology, USA
Anatoli Djanatliev, University of Erlangen-Nuremberg, Germany
Thuy T. Do, Luther College, USA
Alexandre Douplik, Ryerson University /St. Michael Hospital, Canada
António Dourado, University of Coimbra, Portugal
Stephan Dreiseitl, University of Applied Sciences Upper Austria, Austria
Mounîm A. El Yacoubi, Telecom SudParis / Institut Polytechnique de Paris, France
Mahmoud Elbattah, University of the West of England Bristol, UK / Université de Picardie Jules Verne, France
Şahika Eroğlu, Hacettepe University, Ankara, Turkey
Gokce Banu Laleci Erturkmen, SRDC A.S., Turkey
Shayan Fazeli, UCLA, USA
(David) Dagan Feng, University of Sydney, Australia
Ana Isabel Ferreira, Nova School of Science & Technology – NOVA University of Lisbon / Health School – Polytechnic Institute of Beja, Portugal
Filipe Fidalgo, Instituto Politécnico de Castelo Branco, Portugal
Duarte Folgado, Associação Fraunhofer Portugal Research | NOVA School of Science and Technology - LIBPhys-UNL, Portugal
Sebastian Fudickar, Universität Oldenburg, Germany
Rosalba Giugno, University of Verona, Italy
Alexandra González Aguña, University of Alcala, Spain
María Adela Grando, ArizonaState University, USA
David Greenhalgh, University of Strathclyde, UK
Abir Hadriche, ENIS - Sfax University, Tunisia
Muhammad Hasan, Texas A&m International University (TAMIU), USA
Sara Herrero Jaén, University of Alcalá, Spain
Mohamed Hosni, ENSAM | Moulay Ismail University, Meknes, Morocco
Wen-Chen Hu, University of North Dakota, USA
Yan Hu, Blekinge Institute of Technology, Sweden
Fábio Iaione, Universidade Federal de Mato Grosso do Sul, Brazil
Tunazzina Islam, Purdue University, USA

Emanuele Rizzuto, SAPIENZA University of Rome, Italy
Sandra Rua Ventura, Center for Rehabilitation Research |School of Health | Polytechnic of Porto, Portugal
Vangelis Sakkalis, Institute of Computer Science - Foundation for Research and Technology (ICS - FORTH), Greece
Ahmad Salehi, Monash University, Australia
Antonio Salvatore Filograna, Engineering Ingegneria Informatica S.p.A., Italy
Patricia Santos, NOVA School of Science and Technology - NOVA University of Lisbon / Superior School of Health of Polytechnic Institute of Beja, Portugal
Alessandra Scotto di Freca, University of Cassino and Southern Lazio, Italy
Jayanthi Sivaswamy, International Institute of Information Technology (IIIT), Hyderabad, India
Pedro Sousa, Nursing School of Coimbra / Center for Innovative Care and Health Technology, Portugal
Arun Sundararaman, Accenture Technology, India
Zoltán Szlávik, myTomorrows, Netherlands
Toshiyo Tamura, Waseda University, Japan
Adel Taweel, Birzeit University, PS/ King's College London, UK
Rafika Thabet, Grenoble-Alpes | INP | CNRS | G-SCOP, France
Ljiljana Trajkovic, Simon Fraser University, Canada
Tuan Tran, College of Pharmacy | California Northstate University, USA
Athanasios Tsanas, University of Edinburgh, UK
Edward Tsien, University of South Carolina / Raven Risk AI, USA
Manolis Tsiknakis, Hellenic Mediteranean University / Foundation for Research and Technology Hellas (FORTH), Greece
Ioan Tudosa, University of Sannio, Italy
Jonathan Turner, Technological University Dublin, Ireland
Gary Ushaw, Newcastle University, UK
Ali Valehi, University of Southern California, USA
Maria Vasconcelos, Fraunhofer Portugal AICOS, Portugal
Agnes Vathy-Fogarassy, University of Pannonia, Hungary
Enrico Vicario, University of Florence, Italy
João L. Vilaça, 2Ai - School of Technology | IPCA, Barcelos, Portugal
Klemens Waldhör, FOM Hochschulzentrum Nürnberg, Germany
Shin'ichi Warisawa, The University of Tokyo, Japan
Pengcheng Xi, National Research Council of Canada / University of Waterloo, Canada
Zongxing Xie, Stony Brook University, USA
Sule Yildirim-Yayilgan, Norwegian University of Science & Technology, Norway
Malik Yousef, Zefat Academic College | Galilee Digital Health Research Center (GDH), Israel
Bing Zhou, Snap Research, USA
Stelios Zimeras, University of the Aegean, Greece

**Copyright Information**

For your reference, this is the text governing the copyright release for material published by IARIA.

The copyright release is a transfer of publication rights, which allows IARIA and its partners to drive the dissemination of the published material. This allows IARIA to give articles increased visibility via distribution, inclusion in libraries, and arrangements for submission to indexes.

I, the undersigned, declare that the article is original, and that I represent the authors of this article in the copyright release matters. If this work has been done as work-for-hire, I have obtained all necessary clearances to execute a copyright release. I hereby irrevocably transfer exclusive copyright for this material to IARIA. I give IARIA permission or reproduce the work in any media format such as, but not limited to, print, digital, or electronic. I give IARIA permission to distribute the materials without restriction to any institutions or individuals. I give IARIA permission to submit the work for inclusion in article repositories as IARIA sees fit.

I, the undersigned, declare that to the best of my knowledge, the article is does not contain libelous or otherwise unlawful contents or invading the right of privacy or infringing on a proprietary right.

Following the copyright release, any circulated version of the article must bear the copyright notice and any header and footer information that IARIA applies to the published article.

IARIA grants royalty-free permission to the authors to disseminate the work, under the above provisions, for any academic, commercial, or industrial use. IARIA grants royalty-free permission to any individuals or institutions to make the article available electronically, online, or in print.

IARIA acknowledges that rights to any algorithm, process, procedure, apparatus, or articles of manufacture remain with the authors and their employers.

I, the undersigned, understand that IARIA will not be liable, in contract, tort (including, without limitation, negligence), pre-contract or other representations (other than fraudulent misrepresentations) or otherwise in connection with the publication of my work.

Exception to the above is made for work-for-hire performed while employed by the government. In that case, copyright to the material remains with the said government. The rightful owners (authors and government entity) grant unlimited and unrestricted permission to IARIA, IARIA's contractors, and IARIA's partners to further distribute the work.

# Real-world ADL Recognition with Deep Learning and Smartwatches: A Pilot Study

Mustafa Ahmed, Pelagia Koufaki
Centre for Health, Activity and Rehabilitation Research
Queen Margaret University,
Edinburgh, UK
e-mail: Mahmed@qmu.ac.uk,
Pkoufaki@qmu.ac.uk

Hongnian Yu
School of Engineering and the Built Environment
Edinburgh Napier University,
Edinburgh, UK
e-mail: H.Yu@napier.ac.uk

Michael Vassallo
Department of Medicine for Older People
Royal Bournemouth Hospital,
Bournemouth, UK
e-mail: michael.vassallo@uhd.nhs.uk

*Abstract*— **The global aging population poses significant challenges to healthcare systems, especially in promoting independent living and reducing caregiver burdens. Technology-Enabled Care (TEC), which leverages digital tools and Artificial Intelligence (AI), has emerged as a promising solution to support older adults. A crucial component within TEC is the automatic recognition of Activities of Daily Living (ADLs), essential for early detection of health declines and personalized care. Traditional ADL recognition research, often conducted in controlled environments, does not adequately address real-world complexities. This study bridges the gap between laboratory prototypes and practical applications by developing a user-friendly ADL recognition framework using commercial smartwatches. A hybrid model, combining Convolutional Neural Networks (CNN) and Long Short-Term Memory (LSTM) networks, was trained on accelerometer and gyroscope data to recognize activities like dishwashing and walking. Initially validated in a lab setting with an accuracy of 94%, the model was subsequently tested over a 20-day pilot study involving five participants (mean age = 32 years, SD = 4.5), each wearing an Apple Watch device. Real-world results revealed a significant performance drop, with accuracy declining to 81%. Activities like mopping maintained high recognition accuracy, while subtler tasks, such as walking and washing face posed challenges due to movement variability. These findings underscore the need for model optimization using real-world data to improve recognition accuracy and address variability in movement patterns. Further research is essential to refine these systems for broader applications, develop strategies to enhance user adherence, and ultimately support the independence and well-being of aging individuals.**

***Keywords-Activities of Daily Living; Activity Recognition; Deep Learning; Independent living; Smartwatch.***

## I. INTRODUCTION

The global population is aging rapidly, with a projected 1.5 billion individuals exceeding 65 years old by 2050 [1]. This demographic shift strains healthcare systems as older adults experience higher rates of chronic conditions and functional limitations [2]. Accurate assessment of functional health is crucial for early detection of decline, enabling timely interventions and improved quality of life [3]. In this regard, Technology-Enabled Care (TEC) offers significant advantages over traditional self-reported and clinical observation methods for functional assessment [4]. By providing continuous, objective, and comprehensive monitoring, it facilitates early detection of health issues, timely interventions, and personalized care plans, ultimately enhancing the quality of life and independence of older adults.

Automatic recognition of Activities of Daily Living (ADLs) is a key area within TEC. ADLs, such as brushing teeth, washing dishes, and cleaning the house, are fundamental for independent living and serve as indicators of an individual's functional health [5]. Identifying subtle changes in ADL performance potentially allows for preventative measures and interventions before decline becomes significant [6]. Efficient ADL Recognition (ADL-R) systems can offer users valuable insights into their daily activities, helping them improve routines and adopt healthier behaviors [7]. For caregivers, these systems enhance understanding of the care recipient's needs and patterns, leading to more effective and responsive caregiving and informed decision-making [8]. Additionally, clinicians can remotely monitor patients, thereby reducing the need for frequent in-person visits and facilitating more efficient management of chronic conditions. [9] This allows for timely interventions that can prevent hospitalizations.

Wearable technology and AI have made significant strides in healthcare monitoring, enabling continuous, multimodal assessments that provide a comprehensive view of health. Recent innovations, such as hybrid sensors, track both biochemical and biophysical signals, offering more detailed insights compared to single-parameter devices [10]. Fiber-based strain sensors have also contributed by enhancing flexibility and diagnostic capabilities, while reducing costs, making wearable devices more practical and accessible [11]. These advancements have also expanded the potential of wearable devices in Human Activity Recognition (HAR) and ADL-R. Inertial sensors like accelerometers and gyroscopes are increasingly used as privacy-conscious alternatives to camera-based systems, offering reliable, continuous monitoring that suits personal and home environments [12]. As a result, wearables are becoming a valuable tool for ADL-

R in healthcare, where they can provide insights into routines and health conditions in real time.

However, several challenges hinder the widespread adoption of ADL-R systems outside controlled laboratory environments. Many wearables designed for research are bulky and uncomfortable, making them impractical for daily wear [13]. The complex machine learning models necessary for accurate activity recognition can strain device resources, leading to frequent charging requirements and technical difficulties [14]. Real-life deployments also face logistical hurdles, such as ensuring users can operate the devices independently and maintaining consistency across settings, which often require home visits and substantial user training [15]. These issues highlight a broader limitation of existing solutions: while they may excel in specific contexts, such as tracking upper limb movements for rehabilitation using deep learning models [16], they lack the versatility needed for broader ADL applications. Additionally, stationary sensor setups, like those integrating Wi-Fi and Inertial Measurement Unit (IMU) data [17], though promising for capturing detailed activity characteristics, are often impractical for mobile and daily use.

To address these challenges, recent developments in wearable technology focus on improving user-friendliness and efficiency. For instance, wearable devices that combine photoplethysmography and inertial data offer a more holistic assessment by capturing a broader range of signals [18], though they may introduce added complexity and user burden. Adaptive algorithms, such as "one-size-fits-most" models, generalize across devices and body locations, enhancing accuracy for specific activities like walking [19]. However, these models often struggle to account for the full diversity of daily activities in natural settings.

A user-centric approach that leverages familiar, widely used devices like smartwatches and smartphones offers a promising solution for real-world ADL-R. These devices are accessible, comfortable, and capable of supporting ADL-R systems that optimize for low power consumption, reducing the need for frequent charging and improving practicality for long-term use. Recent studies emphasize that optimizing AI models specifically for these devices not only extends battery life but also ensures that ADL-R is sustainable and adaptable to everyday environments [20].

This pilot study specifically aims to evaluate the feasibility of a smartwatch-based framework in addressing ADL-R challenges under real-life conditions. By using a single, widely adopted smartwatch, we address the common limitations of bulkiness and impracticality that have plagued previous ADL-R systems, enabling a more accessible and minimally intrusive approach to recognize ADLs in natural, home-based environments. This user-centric design reduces the need for specialized equipment and provides a sustainable solution for real-world health monitoring that overcomes these significant barriers. By emphasizing user-friendliness and computational efficiency, this pilot study lays the groundwork for developing robust, accessible TEC solutions for ADL-R. Through this feasibility assessment, we aim to pave the way for broader adoption and improved functional health monitoring, particularly benefiting aging populations who require practical and scalable ADL-R systems.

The remainder of this paper is organized as follows: Section II outlines the methods, including model design, participant recruitment, and data collection procedures. Section III presents the results, covering model performance and participant engagement. Section IV discusses the findings, addressing discrepancies and engagement factors. Section V explores implications for future research and advancements in ADL-R.

## II. Methods

### *A. ADL Recognition Model*

The activity recognition model was designed as a hybrid architecture combining lightweight Convolutional Neural Networks (CNN) and Long Short-Term Memory (LSTM) networks to leverage the strengths of both approaches in handling sensor data. The model was trained on accelerometer and gyroscope data collected in a simulated living environment, focusing on target activities, which were: dishwashing, shelving items, brushing teeth, washing face, mopping & hoovering, and walking. Data was sampled at 34Hz and segmented into 442-sample windows, approximately 13 seconds each. The CNN part of the model consisted of multiple convolutional layers, each with varying filter sizes and numbers of filters to extract spatial features from the raw sensor data. Typically, the first convolutional layer used 32 filters with a 3x1 kernel size and Rectified Linear Unit (ReLU) activation, followed by a second convolutional layer with 64 filters of the same kernel size and activation function. Max-pooling layers were used after certain convolutional layers to reduce the dimensionality of the feature maps while retaining important spatial features. The final convolutional layer was followed by a flatten layer, converting the 2D feature maps into a 1D feature vector for the LSTM layers. The LSTM network then processed this feature vector to capture temporal dependencies and sequential patterns in the activity data. Typical configurations included an initial LSTM layer with 128 units and a dropout rate of 0.2, followed by a second LSTM layer with 64 units and the same dropout rate. A grid search was conducted to optimize hyperparameters for both the CNN and LSTM components, including the learning rate, batch size, and dropout rate. The learning rate was tested in the range of 0.001 to 0.01, with 0.001 selected. Batch sizes of 16, 32, and 64 were evaluated, and 32 was chosen. Dropout rates from 0.2 to 0.5 were assessed specifically for the LSTM layer, with 0.2 providing optimal regularization. The output layer was a fully connected dense layer with a SoftMax activation function to predict the activity classes.

The model was trained using the Adam optimizer with categorical cross-entropy as the loss function and then evaluated using a 5-fold cross-validation approach to ensure an unbiased assessment of its performance. Accuracy and F1-score were used to evaluate the model's effectiveness in recognizing the target activities. After the model was developed and initially tested, a pilot study was designed to validate its feasibility and effectiveness in real-life settings.

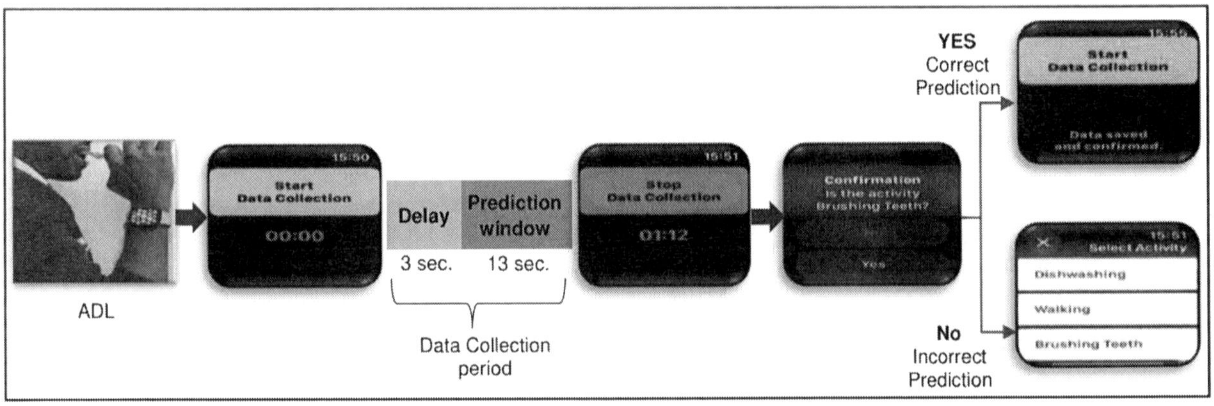


Figure 1. Workflow of the smartwatch app's data collection and activity prediction process.

This study aimed to assess the practical application of the ADL-R system, focusing on user interaction, data collection, and system performance in everyday environments. Over the course of 20 consecutive days, participants wore Apple watch devices (series 8) that collected motion data through a custom-developed application and used these data to recognize the activity performed.

### B. *Participants*

Participants were recruited through convenience sampling, primarily targeting individuals within the University community for convenience and accessibility. Five participants (mean age = 32 years, SD = 4.5), including two females and three males, were enrolled. The only inclusion criterion was to have access to an iPhone. All participants were generally familiar with technology and provided written informed consent before commencement of data collection. The study was reviewed and received a favorable ethical opinion by the Queen Margaret University Ethics Committee, (REP 0278).

### C. *Real-life ADL Recognition procedures*

The custom-developed iOS application, QMU ADL Tracker, was created using Xcode and deployed via Apple TestFlight for easy installation and updates. The application consists of a smartphone app with a companion smartwatch app, designed to facilitate seamless data collection and user interaction. Data collection is conducted through the CoreMotion framework, ensuring consistent and precise capture of sensor data. The CNN-LSTM model was deployed in the app through CreateML.

The smartwatch app features a user-friendly interface (Fig. 1), including a Start/Stop toggle button that simplifies the process of beginning and ending data collection. To ensure accurate activity recognition, users are instructed to start data collection prior to performing an activity and to stop it afterward. Once data collection is stopped, the sensor data is processed by the CNN-LSTM model, which provides a real-time prediction of the activity. Users are then prompted to confirm the prediction's accuracy by selecting "Yes" or "No" (Fig. 1). If the prediction is correct, users press "Yes." If incorrect, they press "No," prompting a list of remaining target ADLs for selection. In cases where the user's activity is not a target ADL, an "Other" button allows access to a list of additional activities, such as tidying up, cleaning windows, driving, shopping, sitting, lying down, eating, drinking, and preparing meals. Once the user confirms the activity, both the motion data and the user's selection are sent to the smartphone app for storage. The smartphone app primarily functions as a data repository, allowing users to view collected information and providing instructional support.

Data collection parameters, including sampling frequency and window size, were aligned with those used in the simulated environment for consistency with the CNN-LSTM model. As shown in Fig. 1, the application was configured to include a 3-second buffer period to stabilize sensor readings before data collection begins. Consequently, the minimum total time required for a prediction was 16 seconds (3 seconds delay plus a 13-second prediction window). Participants were instructed to perform each activity for at least 16 seconds to ensure accurate predictions.

## III. RESULTS

### A. *Model Performance: Simulated Data Training and Testing*

As shown in Table I, the validation of the CNN-LSTM model assessed accuracy and F1 score for each activity, with the F1 score reflecting the balance between precision and recall, indicating the model's accuracy in classifying true positives from false positives and negatives. The model achieved an overall accuracy of 94% and an F1-score of 93% in recognizing ADLs within the simulated environment. A more granular analysis reveals that shelving items exhibited the highest accuracy (99%) and respectable F1-score (94%), suggesting robust recognition of this activity. Conversely, while washing face achieved a high accuracy of 96%, its F1-score of 89% indicated potential challenges in correctly identifying this activity. The model's confusion matrix, as shown in Fig. 2, provides a detailed visualization of the model's predictions compared to the true activities. To further

TABLE I. CNN- LSTM MODEL PERFROMANCE ON REAL-LIFE DATASET

| Activity | Accuracy | F1 Score |
|---|---|---|
| Walking | 95% | 91% |
| Brushing Teeth | 95.8% | 85.6% |
| Washing Face | 96% | 79% |
| Mopping/Hoovering | 97% | 94.7% |
| Dishwashing | 92.5% | 80.9% |
| Shelving Items | 99% | 94.5% |
| **Overall** | **94%** | **93%** |

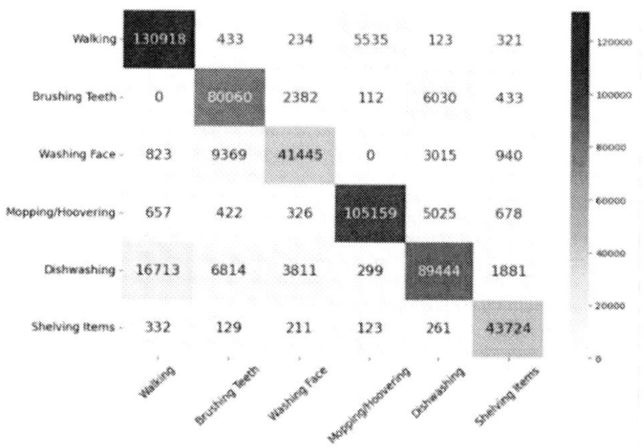


Figure 2. Confusion Matrix for Model's predictions using simulated data.

TABLE II. CNN- LSTM MODEL PERFROMANCE ON REAL-LIFE DATASET

| Activity | Accuracy | F1 Score |
|---|---|---|
| Walking | 68% | 68% |
| Brushing Teeth | 85.5% | 85.5% |
| Washing Face | 65.8% | 45.8% |
| Mopping/Hoovering | 95% | 95% |
| Dishwashing | 79% | 72.5% |
| Shelving Items | 92.6% | 92.6% |
| **Overall** | **81%** | **76.5%** |

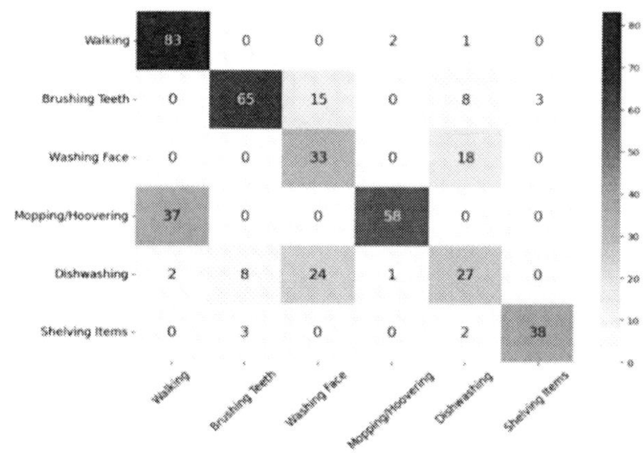


Figure 3. Confusion Matrix for Model's predictions using simulated data.

investigate the model's generalization capabilities, we carried out the validation in real-life through this pilot study.

## *B. Model Performanece in real-life*

Feedback from participants on the predicted activities provided valuable insights into the model's performance. Based on this feedback, we constructed a confusion matrix (Fig. 3) to quantitatively assess the accuracy of activity recognition. The matrix revealed an overall accuracy of 81% in identifying daily activities in real-world settings, indicating a moderate level of success in ADL-R. However, this also highlights the challenges of applying the model to real-life scenarios.

As detailed in Table II, the activities of mopping/hoovering achieved the highest accuracy at 95.08%, which suggests that the model is particularly adept at recognizing the distinct motion patterns associated with this activity. Despite its high accuracy, mopping/hoovering was sometimes misclassified as dishwashing. Another activity, shelving items, showed a similarly commendable accuracy of 92.68%. While it was generally well recognized, there were instances of confusion with the activity of brushing teeth, likely due to the similar repetitive hand movements involved. The high accuracy rates for these specific activities demonstrate the model's effectiveness in distinguishing certain types of ADLs.

On the other hand, washing face had the lowest accuracy at 65.83%, with frequent misclassifications as brushing teeth and dishwashing. This high confusion rate highlights the challenge in recognizing the movements involved in washing the face, pointing to a need for better feature differentiation and potentially additional sensor data. Unexpectedly, walking also exhibited a low accuracy of 68.21%, with the primary confusion occurring with mopping/hoovering. The substantial misclassification rate indicates that the model struggles to distinguish between these activities, possibly due to similar sensor data patterns. Brushing teeth exhibited a moderate accuracy of 85.53%. Misclassifications primarily occurred with dishwashing and shelving items, suggesting challenges in distinguishing repetitive hand movements across these tasks. Similarly, dishwashing had a moderate accuracy of 79.03%, often confused with brushing teeth and washing face. This overlap in recognition points to the difficulty in differentiating between activities involving similar hand and arm movements, indicating a need for refined feature extraction to improve accuracy.

## *C. Participants' Engagement*

Various indicators related to participant engagement with the ADL tracker app are shown in Fig. 4, which reveals variations in how participants used the app. In Fig. 4A, the number of each activity participants performed using the application during the study data collection period is illustrated. On average, participants logged approximately 81 activities over the 20-day period, with noticeable individual differences in activity logging. Participant P5 exhibited the highest level of engagement, recording 115 activities, while Participant P2 logged the fewest with 61.

The frequency of app usage, as measured by average sessions per day (Fig. 4B), was consistent across participant with a mean of four sessions. This suggests a limited engagement with the study's requirements. Nevertheless, the

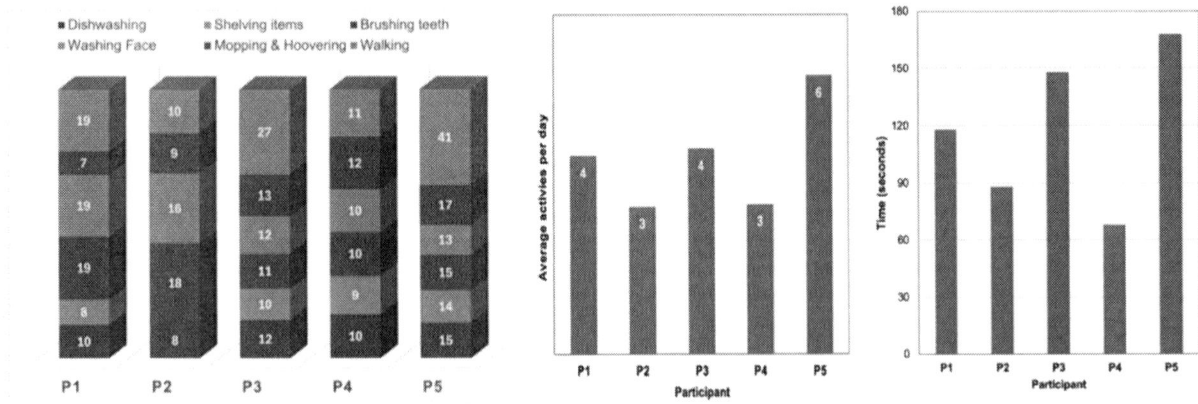


(A) (B) (C)

Figure 4. Indicators of participants engagement showing: (A): Number of each ADL logged by participants during the 20-day data collection period, (B): Frequency of app usage (sessions per day), and (C): Average duration of data collection in each session.

duration of these sessions, as shown in Fig. 4C, varied considerably. Participant P3 demonstrated the longest average session duration at 148 seconds, while Participant P4 recorded the shortest at 68 seconds. These findings imply differences in how participants utilized the app, with some spending longer periods per session, potentially logging multiple activities or engaging in more detailed data exploration.

The analysis of the activities performed by participants reveals several important trends and patterns. Notably, walking was the most frequently performed activity, with participants engaging in it between 10 and 41 times over the testing period. Additionally, personal hygiene activities such as brushing teeth and washing face were consistently performed by all participants, indicating regular adherence to daily hygiene routines. Less frequent activities, such as mopping & hoovering and shelving items, were performed less often compared to other activities. This lower frequency could be attributed to the nature of these activities, which may not occur on a daily basis, resulting in fewer recorded instances.

Examining participant-specific trends, Participant P5 demonstrated the highest overall engagement, particularly in walking (41 instances) and dishwashing (15 instances). Conversely, Participant P2 did not record any instances of shelving items, which may indicate a lack of engagement in this specific activity or a potential oversight in the recording. The data also reflects a diverse range of activities performed by the participants, with each individual engaging in multiple types of activities. Participant P4, for example, exhibited a balanced engagement across all activities, performing at least nine instances of each, which is beneficial for comprehensive training of the recognition model.

## IV. Discussion

### A. *Model Performance and the Gap Between Simulation and Reality*

The proposed hybrid CNN-LSTM model demonstrated good performance in the simulated environment, achieving a notable accuracy of 94% in ADL-R. This highlights the potential of such architectures in handling sensor data for activity recognition tasks. However, the stark contrast between simulated and real-world performance, with an overall accuracy of 81% in the latter, underscores the challenges inherent in bridging the gap between controlled environments and complex, dynamic real-world settings.

Several factors may contribute to this performance discrepancy. The simulated environment likely presents a more idealized representation of ADLs, with controlled conditions and limited variability in sensor data. In contrast, real-world activities are subject to a multitude of factors, including environmental noise, variations in how objects are utilized, and the inherent variability of wearable sensor performance. These complexities introduce significant challenges for the model, hindering its ability to generalize effectively.

Moreover, the way participants performed ADLs in the real world may have differed substantially from the simulated patterns. The model, trained on simulated data, might not have been adequately prepared to handle the diverse and nuanced variations observed in real-life behavior. This discrepancy highlights the need for more representative training data that captures the full spectrum of human activity.

The differential performance of the model across different ADLs provides valuable insights into the factors influencing recognition accuracy. Activities like mopping/hoovering and shelving items, characterized by distinct and repetitive movement patterns, were recognized with high accuracy. This suggests that the model can effectively capture and classify well-defined activities. In contrast, washing face and walking presented significant challenges. The low accuracy for washing face might be attributed to the subtle and often occluded movements involved in this activity, making it difficult to differentiate from similar actions. For walking, the confusion with mopping/hoovering suggests potential overlap in sensor patterns, especially when considering variations in walking speed and style.

These results highlight the complexities involved in automatic feature extraction. While the model's ability to learn discriminative features directly from raw sensor data is advantageous, it also presents certain limitations. A hybrid approach, combining automatically learned features with carefully crafted domain-specific features, could offer a promising avenue for enhancing the model's overall performance. By leveraging the strengths of both approaches, it may be possible to address the challenges posed by complex and varied ADLs.

### *B. Participant Engagement and Data Quality*

Participant engagement in the ADL tracker app varied significantly, influencing the quantity and quality of data collected, which, in turn, impacted the model's performance. This variability highlights a critical challenge: while some participants frequently interacted with the app, others engaged less consistently. Such differences in engagement can affect the representativeness of the dataset, as well as the accuracy and generalizability of the model in real-world settings.

Several factors may contribute to these engagement disparities. One primary challenge is that certain ADLs, like mopping/hoovering or shelving items, are not performed frequently, which naturally leads to less frequent app usage. Additionally, the perceived inconvenience of wearing a device throughout the day and a lack of immediate, visible benefits from using the app may also contribute to lower engagement levels. These factors underscore the difficulty of integrating wearable technology seamlessly into everyday life when it doesn't directly align with the user's regular routine. To address these challenges, accurately measuring engagement could provide valuable insights into adherence patterns. Developing metrics that capture not just the frequency of app usage but also the context of interactions would offer a clearer understanding of participant behavior. By gaining a more detailed view of how participants engage with the app, researchers can better align ADL-R models with real-world usage patterns, ultimately enhancing model accuracy and generalizability.

To improve engagement, several strategies could be implemented. Personalized feedback that provides insights into activity patterns can make the data collection process more relevant and motivating for users. Gamification elements, such as rewards for consistent engagement, could foster a sense of accomplishment and incentivize regular app usage. Additionally, in-app reminders may help prompt users to engage without being intrusive. Streamlining the user interface and ensuring that the app operates smoothly in the background could reduce perceived burdens, encouraging participants to incorporate the technology into their routines more naturally. Collectively, these strategies aim to create a more engaging and user-friendly experience, enhancing adherence and supporting the broader application of wearable ADL-R systems.

## V. Implication for Future Research

This pilot study provides first critical insights into the challenges and opportunities for improving ADL-R in real-world settings. It successfully demonstrates the feasibility of conducting ADL-R research using a user-centric approach, highlighting practical applications and potential enhancements. The seamless recruitment of participants with minimal interaction, facilitated by online platforms like TestFlight, and the user-friendly design of the app, enhanced the efficiency of the study. Moreover, the dataset collected in this study serves as a valuable resource for refining ADL-R models. By retraining the model on an expanded sample, we can enhance its accuracy and robustness, enabling it to learn from a wider range of activity patterns and variations. To further improve performance, combining automatically learned features with carefully crafted domain-specific features is essential, leveraging the strengths of both data-driven and knowledge-based approaches. Enhancing feature engineering to capture temporal dynamics, contextual information, and activity transitions can further enrich the model's representational power. In addition to that, exploring advanced model architectures, such as attention mechanisms, transformers, or graph-based approaches can unlock the potential for capturing complex dependencies within the data.

To maximize the potential of ADL-R systems, a paramount focus on user engagement and data quality is essential. Implementing strategies to increase user participation and motivation is crucial for the success of such systems. Understanding the most frequently performed ADLs allows for a more targeted approach to app customization. By focusing on core activities, the app can provide relevant features and notifications, enhancing the user experience and encouraging sustained engagement. This interplay between user engagement, data quality, and app personalization is fundamental to the development of robust ADL-R systems.

Moreover, these insights can inform future developments in wearable technology for ADL-R by emphasizing the importance of user-centric design. As wearables evolve, integrating advanced sensing capabilities that capture a broader range of ADLs could enhance the comprehensiveness of ADL-R systems. Furthermore, the development of adaptive algorithms capable of personalizing recognition based on individual usage patterns can make ADL-R systems more responsive to user needs. Such advancements could expand the scope of wearable technology, making it more versatile for different user demographics and environments, ultimately broadening the impact of ADL-R systems beyond specific study settings. By leveraging data on user engagement and preferences, future wearable devices could incorporate enhanced features, such as context-aware prompts and dynamic feedback, which adjust in real-time to optimize user adherence and data quality.

While this pilot study provides valuable insights, the small sample size may limit the generalizability of the findings.

Future research should aim to recruit a larger and more diverse participant pool, potentially by partnering with community organizations or leveraging online recruitment platforms. This approach would enhance the representativeness of the data and further validate the model's effectiveness in broader real-world settings.

## ACKNOWLEDGMENT

Apple Inc. provided the Apple Watch devices for the research. Apple was not involved in the design of the research, nor was it involved in the collection, analysis, or interpretation of the research data, or the content of this or any related publication.

## REFERENCES

[1] World Health Organization, "Progress report on the United Nations decade of healthy ageing, 2021-2023," World Health Organization, 2023. retrieved: June, 2024, from https://www.who.int/publications/i/item/9789240079694

[2] J. Parajuli, D. Berish, and Y. L. Jao, "Chronic conditions and depressive symptoms in older adults: The mediating role of functional limitations," Aging & Mental Health, vol. 25, no. 2, pp. 243-249, 2021.

[3] A. M. Jahan, "Insight into Functional Decline Assessment in Older Adults: A Physiotherapist's Perspective," Archives of Gerontology and Geriatrics Plus, vol. 100048, 2024.

[4] M. Schmitter-Edgecombe, C. Luna, and D. J. Cook, "Technologies for health assessment, promotion, and intervention: Focus on aging and functional health," in Positive Neuropsychology: Evidence-based perspectives on promoting brain and cognitive health, Cham: Springer International Publishing, pp. 111-138, 2022.

[5] S. Tomsone, V. Horstmann, F. Oswald, and S. Iwarsson, "Aspects of housing and perceived health among ADL independent and ADL dependent groups of older people in three national samples," Aging Clinical and Experimental Research, vol. 25, pp. 317-328, 2013.

[6] Y. Angerova et al., "Utilization of ADL performance tests to predict expected functional status in patients after stroke," KONTAKT-Journal of Nursing & Social Sciences related to Health & Illness, vol. 23, no. 3, pp. 320-326, 2021.

[7] M. J. van Het Bolscher-Niehuis, M. E. den Ouden, H. M. de Vocht, and A. L. Francke, "Effects of self-management support programmes on activities of daily living of older adults: A systematic review," International Journal of Nursing Studies, vol. 61, pp. 230-247, 2016.

[8] S. Hvalič-Touzery and V. Dolničar, "Attitudes towards smart technologies among older people and their informal carers in Slovenia," Health of the Elderly, 2021. Retrieved October, 2024, from https://doi.org/10.26493/978-961-293-129-2.71-80

[9] D. G. Leo et al., "Interactive remote patient monitoring devices for managing chronic health conditions: systematic review and meta-analysis," Journal of Medical Internet Research, vol. 24, no. 11, pp. e35508, 2022.

[10] K. Mahato et al., "Hybrid multimodal wearable sensors for comprehensive health monitoring," Nature Electronics, vol. 1, pp. 1-16, 2024.

[11] J. Zhang et al., "Revolutionizing digital healthcare networks with wearable strain sensors using sustainable fibers," SusMat, vol. e207, 2024.

[12] S. Muhammad, K. Hamza, H. A. Imran, and S. Wazir, "Enhanced Human Activity Recognition Using Inertial Sensor Data from Smart Wearables: A Neural Network Approach with Residual Connections," in 2024 International Conference on Engineering & Computing Technologies (ICECT), IEEE, pp. 1-6, May 2024.

[13] U. Martinez-Hernandez et al., "Wearable assistive robotics: A perspective on current challenges and future trends," Sensors, vol. 21, no. 20, pp. 6751, 2021.

[14] O. El-Gayar and A. Elnoshokaty, "Factors and design features influencing the continued use of wearable devices," Journal of Healthcare Informatics Research, vol. 7, no. 3, pp. 359-385, 2023.

[15] Y. Wang, S. Cang, and H. Yu, "A survey on wearable sensor modality centred human activity recognition in health care," Expert Systems with Applications, vol. 137, pp. 167-190, 2019.

[16] A. S. Nunes, İ. Yildiz Potter, R. K. Mishra, P. Bonato, and A. Vaziri, "A deep learning wearable-based solution for continuous at-home monitoring of upper limb goal-directed movements," Frontiers in Neurology, vol. 14, pp. 1295132, 2024.

[17] W. Guo, S. Yamagishi, and L. Jing, "Human Activity Recognition via Wi-Fi and Inertial Sensors with Machine Learning," IEEE Access, 2024.

[18] N. Hnoohom, S. Mekruksavanich, and A. Jitpattanakul, "Physical activity recognition based on deep learning using photoplethysmography and wearable inertial sensors," Electronics, vol. 12, no. 3, pp. 693, 2023.

[19] M. Straczkiewicz, E. J. Huang, and J. P. Onnela, "A 'one-size-fits-most' walking recognition method for smartphones, smartwatches, and wearable accelerometers," NPJ Digital Medicine, vol. 6, no. 1, pp. 29, 2023.

[20] J. M. Peake, G. Kerr, and J. P. Sullivan, "A critical review of consumer wearables, mobile applications, and equipment for providing biofeedback, monitoring stress, and sleep in physically active populations," Frontiers in Physiology, vol. 9, pp. 743, 2018.

# Challenges and Strategies of Inter-Disciplinary Research and Development: Lessons from a Telehealth Project

Wei Zhou, Andy Li, Rashina Hoda, Chris Bain
*Faculty of Information Technology*
*Monash University*
Melbourne, Australia
e-mail: {wei.zhou2, andy.li, rashina.hoda, chris.a.bain}@monash.edu

Sam Georgy
*Video Call Service,*
*Health Direct Australia,*
Sydney, Australia
e-mail: sam.georgy@healthdirect.org.au

Teresa O'Brien, Mahima Kalla, Olivia Metcalf, Wendy Chapman
*Centre for Digital Transformation of Health*
*University of Melbourne*
Melbourne, Australia
e-mail: {teresa.obrien, mahima.kalla, olivia.metcalf, wendy.chapman}@unimelb.edu.au

Peter Poon
*Supportive and Palliative Care*
*Monash Health*
Melbourne, Australia
e-mail: peter.poon@monash.edu

***Abstract*—The rapid rise of telehealth has revolutionized healthcare access, particularly for those in rural and underserved areas. This paper presents the interdisciplinary research and development of a telehealth project aimed at enhancing the experience of palliative care patients. The project, spanning three years, involved multiple stakeholders, including software engineering researchers, healthcare researchers, clinical experts, a telehealth industry partner, and a government entity. We address the core challenges encountered during the project, particularly in managing interdisciplinary collaboration, and share strategies that we found useful to enable valuable, human-centered outcomes.**

***Keywords-Telehealth; Interdisciplinary; Palliative care; Research and Development (R&D).***

## I. INTRODUCTION

Interdisciplinary team science [1] is becoming increasingly critical for addressing complex challenges that demand input from multiple disciplines. In domains like digital health, where technology, healthcare, and user experience converge, the integration of diverse expertise drives innovation and enables the development of comprehensive, user-centered solutions. A prominent example is telehealth [2], which is part of connected health [3], and has emerged as an essential component of healthcare delivery, particularly in remote areas and especially during and after the COVID-19 pandemic. This modality enables improved accessibility and equity across diverse populations [4]. Its effective implementation, however, requires overcoming challenges related to technology design, clinical integration, and user experience [5].

In a three-year project aimed at enhancing telehealth led by software engineering, an interdisciplinary team of software engineering researchers, healthcare researchers, clinical experts, a telehealth industry partner, and a government entity brought together a wealth of diverse expertise and experiences. Co-design and evaluation with end-users – patients, carers, and healthcare practitioners – provided critical focus and direction. The experience included several challenges in aligning different disciplinary perspectives [6]. This paper presents an overview of the interdisciplinary challenges encountered during the project and the strategies employed to address them. It emphasizes the importance of team-building, co-design methodologies, and an agile research and development approach to facilitating successful interdisciplinary collaboration [7]. By addressing structural and communication barriers inherent in interdisciplinary work, the project highlights how interdisciplinary teams can effectively co-create innovative solutions in complex and rapidly evolving areas.

The rest of this paper is organized as follows. In Section II, we describe the challenges faced during interdisciplinary R&D, such as differences in terminology, conflicting stakeholder interests, and recruitment issues. Section III outlines the strategies used to overcome these challenges, including team-building, co-design with users, and an agile R&D approach. Section IV concludes with a summary of findings and suggests future steps for improving interdisciplinary projects in digital health.

## II. CHALLENGES IN INTER-DISCIPLINARY R&D

### *A. Differences in Terminologies and Protocols*

A significant challenge in interdisciplinary projects is the discrepancy in terminologies, protocols, and conceptual frameworks between the healthcare and technology sectors. These can lead to miscommunication, misunderstandings, and delays in project timelines. For example, healthcare professionals and software developers may use different definitions for "evaluation", leading to confusion about the outcomes of certain stages in the project. Similarly, terms such as "development" and "implementation" and even "co-design", albeit more subtle, can mean different things in the two disciplines.

### *B. Conflicting Interests and Expectations Across Stakeholders*

Interdisciplinary projects often involve stakeholders with diverse and sometimes conflicting goals. Researchers may prioritize scientific contributions, while industry partners focus on product viability and market potential. This conflict of interest can create tension in decision-making processes, requiring careful negotiation and alignment of objectives.

### C. Participant Recruitment Challenges

We faced significant recruitment challenges due to: many potential participants not meeting the inclusion criteria or clinical gate-keeping, patients having limited access to or familiarity with digital tools, ethical complexity of obtaining informed consent from palliative care patients, and busy clinician schedules [8]. The COVID-19 pandemic exacerbated existing challenges, particularly in the recruitment of representative participants for co-design and evaluation. Unprecedented burdens on the healthcare ecosystem meant participating in research projects was not a priority.

### D. Covering the Full R&D Lifecycle

The project covered the full R&D lifecycle – from ideation and co-design with patients, clincians, and carers to iterative software design, development, and evaluation – which required expertise from multiple disciplines. This complexity often leads to coordination overheads, where ensuring all team members were aligned in their efforts and one activity successfully feeding into the other was non-trivial.

## III. Strategies for Interdisciplinary R&D

Several strategies informed by agile principles [7] and the specific demands of the project were implemented:

### A. Team-Building and Open Communication

Building trust and ensuring open communication were critical to overcoming the challenges of interdisciplinary collaboration. The project utilized team-building exercises, including regular meetings and end-of-year retrospectives, to foster a sense of alignment and shared purpose among stakeholders. Reflexive discussions, where team members were encouraged to share their concerns and suggestions, were held regularly to maintain transparency and address emerging issues.

### B. Co-Design with End-Users

Conducting co-design with end-users, such as patients, carers, and clinicians, allowed the team to avoid a "technology push" approach and instead apply a human-centred approach with a focus on addressing real-world telehealth challenges. By involving end-users in the design process, the team ensured that the solutions developed were relevant and practical for potential deployment in clinical settings. Technology options were considered around end-user needs and workflows, assessed for feasibility and technical compatibility with the industry partner's platform, and developed accordingly, signifying a "technology pull" approach.

### C. Agile R&D Approach

The team adopted an agile R&D approach through iterative cycles of *co-design*, *prototyping*, and *evaluation*, which allowed for flexibility in response to changing circumstances, such as the pandemic's impact on recruitment. This iterative process enabled the team to incorporate feedback from end-users and stakeholders continuously, ensuring that the solutions evolved to meet emerging needs [7].

### D. Working Closely with Industry Partners

Collaboration with industry partners, such as Healthdirect Australia and Monash Health, ensured that the project maintained a focus on real-world applications. These stakeholders provided invaluable insights into the market potential of telehealth solutions, guiding the R&D process toward outcomes that were scientifically robust and commercially viable.

## IV. Conclusion and Future Work

The strategies employed in this project led to several key outcomes, including improved communication and collaboration between healthcare and technology professionals that lasted throughout the project. The team's interdisciplinary approach fostered innovation by bringing together diverse perspectives, ultimately leading to the development of telehealth software enhancements that were clinically relevant and technologically aligned with the industry partner's widely adopted technical platform.

Our experiences from the "enhancing telehealth" project demonstrates the importance of interdisciplinary collaboration in digital health R&D. By building a cohesive team, co-designing with end-users, and adopting agile methodologies, the project successfully navigated the challenges of inter-disciplinary collaboration. Future projects in the digital health space should continue to prioritize interdisciplinary approaches to ensure that innovations are both scientifically rigorous and practically applicable.

## Acknowledgment

This project is supported by the Digital Health CRC Limited (DHCRC), Monash University, Healthdirect Australia, Monash Health, University of Melbourne, and Victorian Department of Health. DHCRC is funded under the Australian Commonwealth's Cooperative Research Centres (CRC) Program.

## References

[1] D. Stokols, K. L. Hall, B. K. Taylor, and R. P. Moser, "The science of team science: Overview of the field and introduction to the supplement," *American journal of preventive medicine*, vol. 35, no. 2, S77–S89, 2008.

[2] E. R. Dorsey and E. J. Topol, "State of telehealth," *New England journal of medicine*, vol. 375, no. 2, pp. 154–161, 2016.

[3] S. T. MacMahon and I. Richardson, "Pathways, technology and the patient—connected health through the lifecycle," *Frontiers in Digital Health*, vol. 5, p. 1 057 518, 2023.

[4] E. Monaghesh and A. Hajizadeh, "The role of telehealth during covid-19 outbreak: A systematic review based on current evidence," *BMC public health*, vol. 20, pp. 1–9, 2020.

[5] S. N. Gajarawala and J. N. Pelkowski, "Telehealth benefits and barriers," *The Journal for Nurse Practitioners*, vol. 17, no. 2, pp. 218–221, 2021.

[6] J. K. Tebes and N. D. Thai, "Interdisciplinary team science and the public: Steps toward a participatory team science.," *American Psychologist*, vol. 73, no. 4, p. 549, 2018.

[7] R. Hoda, N. Salleh, and J. Grundy, "The rise and evolution of agile software development," *IEEE software*, vol. 35, no. 5, pp. 58–63, 2018.
[8] W. Zhou *et al.*, "Recruiting participants in digital health - lessons from a telehealth enhancement project," in *Workshop on Addressing Challenges in Recruiting Participants, IEEE Symposium on Visual Languages and Human-Centric Computing*, Liverpool, UK, 2024, pp. 1–7.

# Hip Fracture Patient Pathways and Agent-based Modelling

Alison N. O'Connor, Stephen E. Ryan, Gauri Vaidya, Paul Harford, and Meghana Kshirsagar,
Department of Computer Science
University of Limerick
Limerick, Ireland
e-mail: {alison.oconnor | stephen.e.ryan | gauri.vaidya | meghana.kshirsagar}@ul.ie
Paul Harford | 23190973@studentmail.ul.ie

***Abstract*—Increased healthcare demand due to ageing populations is significantly straining European services. In Ireland, patients presenting in Emergency Departments with hip fracture injuries should be admitted to Orthopaedic wards or Theatre within four hours. However, the Irish Hip Fracture Database reports that less than 25 % of patients experience this pathway. Digital solutions, including advanced modelling techniques, offer a promising solution to optimising patient flow without impacting day-to-day healthcare provision. In this work, we outline ongoing research that aims to optimise healthcare provision for hip fracture patients through improved data integration and agent-based simulations. We demonstrate how this technology, through enhanced connectivity, can improve healthcare service for hip fracture patients.**

***Keywords-machine learning; patient flow; hip fracture.***

## I. Introduction

Several key factors contribute to recurring overcrowding in Irish hospitals, many of which reflect deep-rooted systemic issues. A major issue driving increased healthcare demand is the complex medical needs and increased Emergency Department (ED) attendance frequency required for older patients (patients aged $\geq 65$ years). Ireland's rapidly ageing population directly correlates to increased patient demand [1], and the European Union (EU) notes a similar trend [2][3]. Data from 2022 paints a stark picture of the severity of the problem. In 2022, over 70% of Irish EDs exceeded their maximum capacity, leaving staff and resources perilously stretched [4]. Even more concerning, 80% of these hospitals reported significant staffing shortages, further compounding the difficulties in managing patient care effectively [5][6]. The shortage of trained healthcare workers has been a persistent issue across the EU [7]. We are now faced with healthcare environments increasingly unable to meet demand [8].

In response to healthcare challenges, Ireland's regulatory body issued a series of recommendations emphasising the importance of effective workforce planning [9]. Digital technologies offer a promising solution to the challenge of resource optimisation in dynamic environments, however, these efforts must be paired with forward-thinking strategies to prepare for future challenges. More recent literature has demonstrated a strong appetite within Ireland for healthcare digital technologies [10]. While Ireland has committed to digitising healthcare, the Irish health ecosystem lags far behind European counterparts [11], making it difficult to ascertain what can be done with existing healthcare datasets. Without decisive action on how we can best use current digital records, the overcrowding crisis will persist, with dire societal consequences. The work focuses on how existing Irish healthcare data can be used to optimise resources.

Resource optimisation is not a new concept, for example, 'Lean manufacturing' is a systems engineering approach aimed at increasing operational efficiency through waste minimisation. It has been highly successful in manufacturing environments [12] and has been adapted to other areas like supply chain management [13] and healthcare [14]. Lean application is process specific and its successful implementation is strongly linked with leadership and company culture [15]. The case-by-case implementation to different hospitals, with different resources, and indeed different cultures, means that it is almost impossible to apply a specific lean method across all environments. These challenges mean that Lean processes are, generally, applied to specific (but often reoccurring) sub-processes. Hence, rather than the large-scale full process implementation we see in manufacturing environments, lean processes in healthcare are not only limited in scope but also cannot be generalised to other locations. These limitations are exacerbated by the fact that lean systems tend to perform poorly when deployed to dynamic processes [16].

We propose that Artificial Intelligence (AI) and Machine Learning (ML) can be used to leverage existing data beyond its original collection purposes. These tools, which have proven abilities to resolve highly non-linear (i.e., dynamic) relationships, can be deployed to bypass the limitations of Lean methods [17]. The goal of this research is to improve patient flow for hip fracture patients through resource optimisation via agent-based modelling. The increased understanding of resource requirements will enable healthcare staff to identify and propose targeted interventions that can be deployed in the real-world. Targeted interventions, deployed in an incremental fashion, can reduce disruption within the acute care setting while also providing adequate feedback on algorithmic performance.

The remainder of this paper is structured as follows: In Section II we introduce Artificial Intelligence (AI) in the context of healthcare services and address why these tools are considered critical to future healthcare operational planning. We also briefly discuss limiting factors surrounding the use of AI tools in healthcare. Section III describes a framework of AI tools that can be used to improve patient flow for hip fracture patients. The potential impacts of this research are addressed in Section IV, with concluding remarks and future work outlined in Section V.

## II. Artificial Intelligence (AI)

Digital technology is revolutionising healthcare by offering new insights through data analysis. Automation reduces the need for manual coordination between departments, and reduces the administrative burden of healthcare workers [18], enabling them to focus on patient care, thus increasing the overall efficiency of the hospital [19]. ML is a sub-field of AI that addresses the methods used by AI to perform specific tasks [20]. ML algorithms have demonstrated abilities in predictive analytics and resource optimisation, which can be used to improve operational efficiency [21]. One of the most significant benefits of ML for healthcare applications is its ability to leverage existing data beyond the original data collection purposes. In Figure 1, we illustrate how existing data can be assessed to identify appropriate ML models, providing actionable insights that can be used to optimise workflows. The integration of AI/ML has long-term financial benefits for healthcare systems. One example is fewer instances of under-staffing, avoiding the financial strain associated with agency based staff, which cost Ireland in excess of €640 million in 2023 alone [22].

While digital solutions offer the potential for transformative change in healthcare provision, they are not without limitations. One limitation is the lack of transparency in algorithm outputs. Simulation methods that perform well

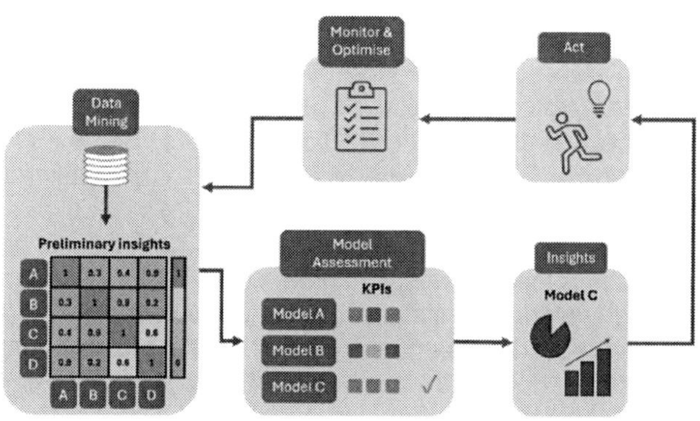


Figure 1. Infographic outlining machine learning model selection and assessment of actionable insights.

on diverse datasets are generally so complex that it is impossible for end-users to interpret model outputs [23]. Enhancing explainability for algorithm outputs remains an active research area [24]. Another limitation is the requirement for large amounts of sensitive personal information, which is highly restricted by Europe's General Data Protection Regulation (GDPR). GDPR has significantly affected researchers' abilities to access meaningful datasets to build, test and/or validate AI/ML models [25]. Despite these limitations, the immense societal benefits of digital health has been recognised at a European level [26]. For hospitals already facing strain due to rising patient numbers and limited resources, these technologies are fast becoming a critical requirement for ensuring that patient care is delivered efficiently without compromising quality. The benefits extend to long-term cost reductions, better resource management, and improvements in patient outcomes, all of which contribute to a more sustainable and effective healthcare system [27].

The question is no longer whether we will implement these technologies, but rather how we can do so effectively and responsibly. In this ongoing collaborative research project, we are investigating the integration of ML technologies into a hospital in the mid-west of Ireland. This initiative aims to explore applications of ML in areas such as predictive patient demand, patient flow optimisation, and resource management, potentially revolutionising healthcare delivery through connected applications. Through this research, we seek not only to enhance patient care and hospital efficiency at a local level but also to establish a framework that could scale nationally, transforming the broader healthcare landscape. Ultimately, the project aspires to position Ireland at the forefront of healthcare innovation, with the potential for these technologies to be adopted on a global scale.

## III. Hip fractures: a case study

The ageing population of Europe has been identified as a significant contributor to overcrowding. Considering the correlation between age and fragility it is not surprising that older people (+65) are at a higher risk of experiencing complex medical fractures [1]. Kanis *et al.* [28] indicates the European cost of fracture services to be in excess of €56 billion (2019) with €290 million attributed to the Irish exchequer. Considering patient flow for a specific sub-set of patients, in this case hip fracture patients, we aim to identify specific causes of poor patient flow enabling targeted resource allocation that impacts both subset and larger population groups. Here, we detail our ongoing research into how digital technologies can be used to improve system operations across an acute care setting for this patient group.

Mid-west Ireland has a population of approximately 500,000 with healthcare provided by the University of Limerick Hospital Group (ULHG). The ULHG comprises six facilities with University Hospital Limerick (UHL) operating as the centralised location for critical care [29]. UHL monitors patients from admission in the ED through the wider acute care setting (e.g., surgical and/or medical procedures) up to patient discharge. Hip fracture patients in the Mid-west requiring longer-term care but with reduced medical needs are often facilitated by other facilities in the UHLG. UHL liaise with the National Office of Clinical Audit (NOCA). NOCA maintains the Irish Hip Fracture Database (IHFD) which collates hip fracture patient information nationally. The purpose of the IHFD is to assess individual hospital compliance against seven Key Performance Indicators (KPIs) including patient time to ward, patient time to the operating theatre [19]. While the overall purpose of the IHFD is to assure quality of healthcare provision, the richness of the data within the IHFD can be exploited far beyond this task. Forecasts predict that hip fracture hospitalisations in Ireland could increase threefold by 2046 [30], which would have a significant impact on Health Service Executive (HSE) resourcing. In 2019, the +65 cohort increased by 1.2% while a 4.2% increase in hip fracture cases was reported [31].

In Figure 2, we illustrate the specific datasets and tools of interest to this work. The outer circle of Figure 2 shows datasets that can be used to provide additional context of value and relevance to hip fracture patients. Immigration, due to global challenges, has significantly increased Ireland's population in recent years, with the population increasing by over 98,000 in the year ending April 2024 [32] alone. The share of elderly ($\geq 65$ years) immigrants has grown in the EU to account for approximately 21% of the immigrant population, and approximately 6% are over 75 years [33]. This has obvious implications for age-related illnesses such as hip fractures, where the mean age of fractures is 83 and 84 years for men and women, respectively. Analysing national census databases [34] enables us to statistically interrogate how national and regional population demographics may affect service demand. The IHFD will provide contextual information related to the average patient age, and the frequency of hip fractures both nationally and regionally.

Most hip fracture patients are posited to arrive via ambulance services but, to our knowledge, this relationship has not been confirmed analytically. Using National Ambulance Service (NAS) records to assess the frequency of NAS requests will enable us to better understand the correlation between this patient group and NAS resource demand. Similarly, previous Stanley *et al.* [35] indicated an association between national weather warnings and hip fractures. However, national weather warnings frequently do not reflect regional weather and the specific type of weather (e.g., rain, ice, snow etc.) was not addressed. Yeung *et al.* [36] however, demonstrates a strong correlation between night-freezing weather events and fall-related injuries.

The middle circle of Figure 2 shows the machine learning algorithms identified as potential tools for analysing data. Time-series analysis will be employed to predict seasonal hip fracture service needs, the average patient length of stay, and the average time spent in EDs and/or surgical units. We previously (Section I) outlined how explainability in AI/ML healthcare is a core requirement for European healthcare services. In this research, we will investigate multiple types of ML algorithms (see Figure 1) and assess their performance across an array of KPIs. One of the KPIs in our research will centre around explainable artificial intelligence, comparing so-called 'black-box' algorithm performance with more interpretable 'white- or glass- box' methods. More advanced simulation methods such as agent-based models, which simulate interaction events, can be used to not only identify the most common patient pathways but also indicate pathway bottlenecks. Agent-based models enable us to interrogate multiple 'what-if' scenarios without impacting day-to-day functioning of acute environments. Test cases of interest to this work include scenarios such as:

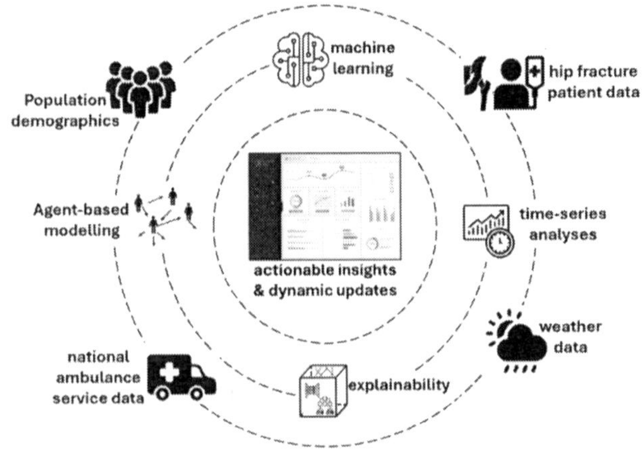


Figure 2. Pipeline of tools investigated in this research.

1) What is the impact of ageing on current service demand, and how will this change in the future based on population changes?
2) What are the optimal staffing resources required to maximise ED service?
3) How do optimal resource requirements change for specific patient groups?
4) What is the average length of stay for hip fracture patients in UHL, and how does this compare with other facilities in the UHLG?

The inner circle of Figure 2 illustrates our expected end goal. Our suite of tools will be used to create a dynamic dashboard highlighting, in an easily interpretable manner, key information related to this patient subgroup. As this work is ongoing, the dashboard display has not yet been fully conceptualised but is likely to include information such as: current number of patients, model prediction of future patients, and the average patient wait time in the ED.

## IV. Impact

In the context of the IHFD, it is interesting to note that data collection frequently occurs post-intervention, which limits the ability of healthcare workers to effect change in a timely manner. Using digital platforms like that illustrated in Figure 2 to automate both data collection and data analysis, opens the possibility of real-time information. Quicker transfer times to surgery for hip fracture patients can reduce the risk of complications such as infections or blood clots, which would otherwise require additional medical interventions and extended hospital stays [19]. Further, this type of system can be integrated with existing hospital data collection systems, reducing the need for manual, repetitive administration. Real-time data access can reduce treatment delays, directly benefiting patients. Accurate ML models can support operational decision-making around workforce planning and resource allocation. Armed with model predictions, hospitals can preemptively adjust staffing levels, prepare operating theatres, and ensure that enough beds are available to meet patient demand. This type of foresight allows hospitals to avoid being overwhelmed, optimises resource use, and prevents operational inefficiencies. By better managing both human and material resources, hospitals can operate more cost-effectively, benefiting not only the institution but also the broader healthcare system. Over time, these improvements contribute to significant cost reductions by improving hospital efficiency and reducing the need for expensive follow-up treatments due to delayed care. Additionally, better resource management means fewer instances of over- or under-staffing, optimizing the use of available personnel and avoiding the financial strain of overtime pay or temporary staff hires during peak periods. The integration of datasets from diverse institutions enhances communication between multiple parties, which can provide an improved patient experience and, more importantly, improved patient outcomes. This research can act as a starting point for larger interoperable systems extending beyond acute care environments to incorporate national and regional information, as outlined here. Future work includes aims to integrate acute- and community- care (e.g., General Practitioners) systems, providing a more holistic healthcare system for Irish citizens.

## V. Conclusion and Future Work

In summary, the integration of digital technology, AI, and ML in healthcare represents a transformative opportunity to address the pressing challenges faced by emergency departments, particularly in the context of the IHFD. The continuous surges in patient volume, especially in emergency settings, have led to significant overcrowding and strain on the healthcare system. By leveraging data-driven insights, hospitals can optimise resource allocation, improve patient flow, and enhance compliance with KPIs. The methods discussed here offer a unique insight into how acute care data collection in Ireland can be leveraged beyond its original purpose, providing actionable insights that can radically improve workflows without negatively impacting daily acute care operations. These technologies not only facilitate the forecasting of patient admissions, but also enhance the prioritisation of care for high-risk patients, ultimately leading to improved patient outcomes. Moreover, the application of AI and ML allows for the continuous optimisation of patient care pathways, ensuring that healthcare resources are used effectively and efficiently.

One significant challenge for this research relates to GDPR restrictions and dataset access. The ethical requirements for healthcare related projects in Ireland are demanding, often requiring researchers in third level institutions to have support from medical staff operating within Ireland's public healthcare system. This can be problematic for researchers in fields like computer science that may not have network connections in medical fields. Ethical applications are generally submitted a month in advance of the committee hearing, and applicants may be invited for in-person review. While committees meet monthly, there are limitations to how many applications can be reviewed at any given time. Even with medical staff support and ethical approval, dataset access is not guaranteed as additional legal requirements around data sharing must be arranged. Most research projects have defined funding timelines, and the slow speed of dataset access presents significant challenges in terms of research outputs. To mitigate these risks, researchers may request anonymised and/or aggregated dataset access. While this eases the ethical application processes, it can have significant implications on ML algorithm precision, especially in terms of prediction capability. Compounding these challenges is the additional limitations on integrating multiple datasets - which could potentially lead to personal data recovery. Together, these limitations not only act to reduce algorithm performance, but also severely limit the potential to uncover relationships between acute- and community-care regimes. While implementing AI/ML technology on restricted and reduced datasets is not ideal, it does offer us the opportunity to identify areas for further exploitation. Preliminary results, generated from these reduced datasets, can be used to showcase the potential of these technologies to improve our healthcare system, which may garner further support for future research works. This project represents an opportunity to test the precision and reliability of explainable AI/ML methods such as:

- Predicting next weeks/months/years' hip fracture cases based on historical accounts and population growth/contraction.

- Classification of patients based on risk profile and treatment requirements (e.g., prioritising high risk patients).
- Clustering analyses to identify patterns in patient movement and resource use to highlight bottlenecks.
- Further patient pathway optimisation through reinforcement learning.

Only by testing these approaches, we can define the limitations and potential uses of these algorithms in supporting healthcare workers.

One future area of research of specific interest is the potential for Radio-Frequency IDentification (RFID) to trace patient pathways from ED arrival through the acute care setting. RFID technology uses radio waves to wirelessly identify object location and is commonly used in postal services to track packages. This technology offers exceptional granular level information about the experienced patient pathway, which is not currently captured by existing healthcare systems. Upon admission, patients with RFID bracelets transmit signals to strategically placed sensors throughout the hospital. These sensors log patient locations and movements, delivering real-time data on each patients' whereabouts within the facility. Analysing the data from RFID sensors allows hospitals to identify the most efficient pathways patients take during their hospital stay. This information can be used to reorganise hospital layouts and workflows, ensuring that patients move smoothly through necessary departments while minimizing delays. Moreover, understanding patient movement patterns enables better resource management, ensuring that staff and equipment are available where and when they are most needed. For instance, if a specific surgical ward consistently experiences delays in patient transfers, this data can prompt an investigation into staffing or logistical issues that may be causing inefficiencies. To effectively communicate findings and insights, data visualization tools can be employed to create dashboards and reports summarizing key metrics and trends. These visualizations help hospital administrators quickly identify areas needing improvement and support data-driven decision-making. By employing these methods, hospitals can harness the power of AI and ML, along with RFID technology, to optimize patient flow, improve compliance with KPIs, and ultimately enhance patient outcomes. This comprehensive approach to data analysis and resource management is essential for addressing the challenges faced by emergency departments and ensuring efficient healthcare delivery in Ireland.

As healthcare systems strive to improve efficiency and patient care, the adoption of these digital technologies and analytical methods will be crucial. The potential for long-term cost savings, improved patient outcomes, and more effective resource management underscores the importance of embracing this technological evolution in healthcare. Ultimately, by harnessing the power of ML hospitals can not only meet current demands but also proactively prepare for future challenges, creating a more resilient and responsive healthcare system for all.

## Acknowledgements

We gratefully acknowledge the expertise, guidance and support of our UHL colleagues Damien Ryan, Carrie Garavan, and Fergal Cummins. This project is funded under the National Challenge Fund which was established under the government's National Recovery and Resilience Plan (NRRP), funded by the EU's Recovery and Resilience Facility. The fund is coordinated and administered by Taighde Éireann – Research Ireland grant 22/NCF/DR/11289.

## References

[1] B. Walsh *et al.*, "Projections of expenditure for primary, community and long-term care in Ireland, 2019-2035, based on the Hippocrates model," The Economic and Social Research Institute and the Minister for Health, Report, Jul. 28, 2021. DOI: 10.26504/rs126.

[2] EUROSTAT, "Ageing Europe - statistics on population developments," Eurostat, 80393, 2020.

[3] EC, *The 2018 Ageing Report: Economic & Budgetary Projections for the 28 EU Member States (2016 2070).* LU: European Commission. Directorate General for Economic and Financial Affairs.Publications Office, 2018.

[4] HIQA, *Emergency Departments Overview Report 2022*, Health Information and Quality Authority (HIQA), 2022.

[5] P. Cullen, "Ireland's population ageing faster than anywhere else in Europe as births fall," The Irish Times, Feb. 12, 2022, [Online]. Available: https://www.irishtimes.com/ireland/social-affairs/2022/12/02/ireland-ageing-faster-than-anywhere-else-in-europe-as-births-fall/ (visited on 03/27/2024).

[6] J. Power, "Pressure on hospitals from Covid surge of 'huge concern' – HSE," The Irish Times, Dec. 27, 2022, [Online]. Available: https://www.irishtimes.com/health/2022/12/27/pressure-on-hospitals-from-covid-surge-of-huge-concern-hse/ (visited on 01/26/2023).

[7] M.-K. Looi, "The European healthcare workforce crisis: How bad is it?" *BMJ*, vol. 384, q8, Jan. 19, 2024, ISSN: 1756-1833. DOI: 10.1136/bmj.q8. pmid: 38242581.

[8] HSE, *HSE National Service Plan 2022*, HSE, Jan. 3, 2022.

[9] HIQA, *Report of the unannounced inspection of the Emergency Department at University Hospital Limerick against the National Standards for Safer Better Healthcare*, Health Information and Quality Authority (HIQA), Mar. 15, 2022.

[10] M. Whelan, "New study shows the public are comfortable with and want to embrace digital technologies in healthcare," HIQA News Updates, Apr. 9, 2024, [Online]. Available: https://www.hiqa.ie/hiqa-news-updates/new-study-shows-public-are-comfortable-and-want-embrace-digital-technologies (visited on 09/26/2024).

[11] DOH, "Minister for Health publishes 'Digital for Care: A Digital Health Framework for Ireland 2024-2030'," May 21, 2024, [Online]. Available: https://www.gov.ie/en/press-release/3ad02-minister-for-health-publishes-digital-for-care-a-digital-health-framework-for-ireland-2024-2030/ (visited on 06/15/2024).

[12] V. Vignesh, M. Suresh, and S. Aramvalarthan, "Lean in service industries: A literature review," *IOP Conference Series: Materials Science and Engineering*, vol. 149, no. 1, p. 012 008, Sep. 2016, ISSN: 1757-899X. DOI: 10.1088/1757-899X/149/1/012008.

[13] J. C. Chen, C.-H. Cheng, and P. B. Huang, "Supply chain management with lean production and RFID application: A case study," *Expert Systems with Applications*, vol. 40, no. 9, pp. 3389–3397, Jul. 1, 2013, ISSN: 0957-4174. DOI: 10.1016/j.eswa.2012.12.047.

[14] L. Rutman, K. Stone, J. Reid, G. A. Woodward, and R. Migita, "Improving Patient Flow Using Lean Methodology: An Emergency Medicine Experience," *Current Treatment Options in Pediatrics*, vol. 1, no. 4, pp. 359–371, Dec. 1, 2015, ISSN: 2198-6088. DOI: 10.1007/s40746-015-0038-0.

[15] G. L. Tortorella, D. Fettermann, F. S. Fogliatto, M. Kumar, and D. Jurburg, "Analysing the influence of organisational culture and leadership styles on the implementation of lean manufacturing," *Production Planning & Control*, vol. 32, no. 15, pp. 1282–1294, Nov. 18, 2021, ISSN: 0953-7287. DOI: 10.1080/09537287.2020.1799255.

[16] A. Azadegan, P. C. Patel, A. Zangoueinezhad, and K. Linderman, "The effect of environmental complexity and environmental dynamism on lean practices," *Journal of Operations Management*, vol. 31, no. 4, pp. 193–212, May 1, 2013, ISSN: 0272-6963. DOI: 10.1016/j.jom.2013.03.002.

[17] M. Shahin, M. Maghanaki, A. Hosseinzadeh, and F. Chen, "Improving operations through a lean AI paradigm: A view to an AI-aided lean manufacturing via versatile convolutional neural network," *The International Journal of Advanced Manufacturing Technology*, vol. 133, no. 11, pp. 5343–5419, Aug. 1, 2024, ISSN: 1433-3015. DOI: 10.1007/s00170-024-13874-4.

[18] H. R. Abbing, "Health, Healthcare and Ageing Populations in Europe, a Human Rights Challenge for European Health Systems Special Issue Section," *European Journal of Health Law*, vol. 23, no. 5, pp. 435–452, 2016.

[19] P. Kelly *et al.*, "Irish Hip Fracture Database National Report 2022," National Office of Clinical Audit (NOCA), 2022, pp. 1–88.

[20] EP, *Artificial Intelligence in Healthcare: Applications, Risks, and Ethical and Societal Impacts.* (Directorate General for Parliamentary Research Services). LU: Publications Office, 2022.

[21] M. M. Rathore, S. Shah, D. Shukla, E. Bentafat, and S. Bakiras, "The Role of AI, Machine Learning, and Big Data in Digital Twinning: A Systematic Literature Review, Challenges, and Opportunities," *IEEE Access*, vol. PP, pp. 1–1, Feb. 22, 2021. DOI: 10.1109/ACCESS.2021.3060863.

[22] J. White, "HSE spends €4.2bn on outside agency staff since 2013," *The Irish Times*, Apr. 6, 2024.

[23] G. Yang, Q. Ye, and J. Xia, "Unbox the black-box for the medical explainable AI via multi-modal and multi-centre data fusion: A mini-review, two showcases and beyond," *An International Journal on Information Fusion*, vol. 77, pp. 29–52, Jan. 2022, ISSN: 1566-2535. DOI: 10.1016/j.inffus.2021.07.016. pmid: 34980946.

[24] S. Ali *et al.*, "Explainable Artificial Intelligence (XAI): What we know and what is left to attain Trustworthy Artificial Intelligence," *Information Fusion*, vol. 99, p. 101 805, Nov. 1, 2023, ISSN: 1566-2535. DOI: 10.1016/j.inffus.2023.101805.

[25] D. Nepelski, *Science for Policy Brief: How can Europe become a global leader in AI in health?* Dec. 2021.

[26] EC, "Artificial Intelligence in Healthcare report | Shaping Europe's digital future," EU Commission, 2021.3613, Nov. 11, 2021.

[27] E. Buijs, E. Maggioni, F. Mazziotta, G. Carrafiello, and F. Lega, *ARTIFICIAL INTELLIGENCE Why and How It Is Revolutionizing Healthcare.* [S.l.]: EMERALD GROUP PUBL, 2024, 1 p., ISBN: 978-1-83549-468-4. DOI: 10.1108/9781835494684.

[28] J. A. Kanis *et al.*, "SCOPE 2021: A new scorecard for osteoporosis in Europe," *Archives of Osteoporosis*, vol. 16, no. 1, p. 82, Jun. 2, 2021, ISSN: 1862-3514. DOI: 10.1007/s11657-020-00871-9.

[29] HSE, "UL Hospitals Group Strategic Plan 2023-2027," UL Hospitals, 2023.

[30] M. A. Kelly *et al.*, "Emerging trends in hospitalisation for fragility fractures in Ireland," *Irish Journal of Medical Science (1971 -)*, vol. 187, no. 3, pp. 601–608, Aug. 1, 2018, ISSN: 1863-4362. DOI: 10.1007/s11845-018-1743-z.

[31] M. Jones, "Improving the Patient Journey for Hip Fracture Patients from Arrival in the Emergency Department to Admission to the Orthopaedic Ward," thesis, Royal College of Surgeons in Ireland, Jul. 27, 2020. DOI: 10.25419/rcsi.12721562.v1.

[32] CSO, "Key Findings Population and Migration Estimates, April 2024 - Central Statistics Office," Population and Migration Estimates, April 2024, Aug. 27, 2024, [Online]. Available: https://www.cso.ie/en/releasesandpublications/ep/p-pme/populationandmigrationestimatesapril2024/keyfindings/ (visited on 09/29/2024).

[33] OECD, "Age distribution in the EU, by citizenship: All ages, 2020," European Commission, Text, 2023. DOI: 10.1787/1d5020a6-en.

[34] CSO, *Central Statistics Office - Publishers - data.gov.ie*, Ireland's open data portal, 2024.

[35] C. Stanley *et al.*, "The association between weather warnings and hip fractures in the Republic of Ireland," *Archives of Osteoporosis*, vol. 18, no. 1, p. 53, Apr. 21, 2023, ISSN: 1862-3514. DOI: 10.1007/s11657-023-01243-9.

[36] M. Yeung, C. Schweitzer, D. Wang, C. Weaver, and E. Lang, "Orthomageddon: A retrospective cohort study of weather-dependent variations in emergency department volume in a Canadian city," *The American Journal of Emergency Medicine*, vol. 38, no. 4, pp. 815–818, Apr. 2020, ISSN: 07356757. DOI: 10.1016/j.ajem.2019.12.009.

# User and Developer Perspectives of Technology Enabled Care

## Initial Findings from Public and Patient Involvement (PPI) and Expert Panel Workshop Studies

Derek Ross
Lero SFI Research Centre for Software
University of Limerick
Limerick, Ireland
Email: derek.ross@ul.ie

Gobe Leinaeng Mphusu
University of Limerick
Limerick, Ireland
Email: 23296224@studentmail.ul.ie

Deepa Narasimha
University of Limerick
Limerick, Ireland
Email: 23138831@studentmail.ul.ie

Kevin Moore
Ei Electronics
Shannon, Ireland
Email: Kevin.Moore@eielectronics.ie

Ann-Marie Morrissey
Ageing Research Centre
Health Research Institute
School of Allied Health
University of Limerick
Limerick, Ireland
Email: Annmarie.morrissey@ul.ie

Ita Richardson
Lero SFI Research Centre for Software
Ageing Research Centre
Health Research Institute
University of Limerick
Limerick, Ireland
Email: Ita.richardson@ul.ie

Katie Crowley
Lero SFI Research Centre for Software
Health Research Institute
University of Limerick
Limerick, Ireland
Email: Katie.crowley@ul.ie

*Abstract*—**Technology Enabled Care (TEC) products and services are expected to be increasingly used by an ageing global population. TEC has the potential to enhance the health and lives of older people. However, factors affecting their acceptance, adoption, and use need to be understood. Two models, the Technology Acceptance Model (TAM) and Health Information Technology Acceptance Model (HITAM) can help to identify acceptance, adoption and use factors. For the current research, two qualitative workshop studies with a Public and Patient Involvement (PPI) panel (N=20), and a TEC Expert group (N=18) were undertaken to identify these factors. The workshops revealed several key issues of importance to the PPI and TEC workshop respondents. These include privacy, costs, and usability of TEC. The use of TAM and HITAM in the workshops proved useful to identifying factors affecting the acceptance, adoption and use TEC by older people.**

***Keywords - Technology Enabled Care; Public and Petient Involvement; Acceptance; Adoption; Use.***

## I. Introduction

Global population ageing is expected to result in an increase in health and related conditions that are associated with age [1]. In order to meet increasing health needs, innovative products and services will be required. Technology Enabled Care (TEC) has the potential to transform the health, well-being and lived experiences of older people. For people living with chronic health conditions and related health risks, health technologies, such as TEC, have the potential to enable continued independent living with minimally invasive or disruptive monitoring and intervention. While healthcare service users are the primary users of TEC, user groups can extend to formal and informal care providers. These secondary users can also benefit from TEC by knowing when and how to provide appropriate care and support. TEC can be used for health monitoring, to support ageing in place, and for prophylactic purposes [2]. Therefore, TEC can provide peace of mind for the various users and stakeholders in a digital health connected system. This suggests a win-win situation whereby TEC can enable people to live better lives into their older age while institutions responsible for care will be better placed to manage the resources required to provide services to older people. Yet, health technology developers and service providers need to be aware of factors that can affect the performance, acceptance and use of the TEC solutions they offer [3][4]. TEC capabilities and limitations need to be well understood to optimise design, development and application in practice. This is the focus of the current research.

Having introduced the rationale for and focus of the current research in Section I, Section II provides an overview of key literature on TEC factors that can affect its acceptance, adoption and use. Following this, Section III states the Research Questions addressed in the empirical work undertaken. Section IV defines the methodology for this research. Sections V and VI describe the context, participants, procedures and findings of the two studies undertaken for this research. In Section VII, the findings are discussed vis-à-vis the literature. Finally, in Section VIII, conclusions drawn and future research proposed.

## II. THE LITERATURE ON TECHNOLOGY ENABLED CARE

The literature on TEC has identified key factors concerning the needs of older people including the provision of healthcare, and the potential for TEC to adequately address these needs.

Two important theoretical contributions to TEC are the Technology Acceptance Model (TAM) [3] and Health Information Technology Acceptance Model (HITAM) [4]. Both TAM and HITAM highlight important factors that underpin acceptance and adoption of TEC. According to TAM, two key factors influence an individuals' willingness to adopt technology: Perceived Usefulness (PU) and Perceived Ease of Use (PEOU) [3]. An extension of TAM, the Patient Technology Acceptance Model focuses on technology acceptance within healthcare settings and incorporates factors including perceived health threats and patient physician relationships. More recently, HITAM has become increasingly recognised as a useful framework [4] for TEC. Unlike TAM, HITAM considers health related aspects, such as health status, beliefs and health information literacy. It has been posited as a valuable framework for studying how older adults embrace technologies that provide health information or aid in making health related decisions [5]. Both TAM and HITAM offer the potential to better understand factors that are key to achieving acceptance and adoption of TEC. In addition, perceived necessity, ease of use and social influences such as the perspective of care providers and peers can help foster positive attitudes and inform older adults' likelihood of acceptance and adoption of TEC [6]-[9].

Although models such as TAM and HITAM offer insights into the factors influencing the acceptance of technology, they may not fully grasp the experiences and requirements of TEC for older adults. The ways in which TEC may impact adults' health, continued independence [10] and overall quality of life need to be addressed [11]. Issues of trust in technology, privacy and the fear of losing independence are significant concerns for TEC use. There may also be organisational and financial obstacles, as well as ethical issues to be considered [12][13]. The preferences of potential TEC users should be considered along with satisfaction and emotional wellbeing in technology impact adoption, use and the potential of positive health outcomes [5][9].

The various issues set out in the literature affords an understanding of factors that can affect the acceptance, adoption and use of TEC by older people. Interdisciplinary insights from gerontology, technology, gerontechnology healthcare, and social sciences can offer an understanding of the complex landscape of TEC for older individuals. Careful attention needs to be paid to these if they are to realise their potential in promoting ageing in place, improving quality of life and reducing caregiver demands.

## III. RESEARCH QUESTIONS

For the current research, it was important to determine an initial understanding of preferences, concerns and functionality features of TEC. To this end, the following research questions were posed:

- What are the actual and/or perceived TEC needs of older people?
- What factors affect the acceptance, adoption and use of TEC by older people?

These were intended to establish target user group and TEC developer perceptions of what product and service features should be incorporated into the design, development and implementation of TEC. They also helped to understand of how well the assumptions of the TEC literature met the perspectives of the user and developer groups.

## IV. METHODOLOGY

### *A) Research Design*

The current research employed a qualitative design that elicited data from two target groups using Workshops as a method. The method is outlined below with the two studies subsequently presented which describe the participants, study specific procedures, and findings. Each study followed a similar Workshop method.

### *B) Workshop Method*

Workshops are a method used to problem solve and or develop new knowledge about an issue of interest [14]. An important contribution of Workshops is that they facilitate genuine participation of target groups and enables researchers to elicit valid and reliable data on specific topics. They are collaborative activities where researchers facilitate the agency of participants to produce new knowledge. Workshops were integral to the research process and data elicitation for the two studies described here. Ørngreen and Levinsen [14] posit the use of Workshops in participatory design and related areas such as Human Centred Design making them particularly relevant to the studies presented below and the interdisciplinary nature of this research.

### *C) Data Analysis*

Data were recorded during Study 1, a Public and Patient Involvement (PPI) Workshop, as field notes by members of the research team, and from data provided by respondents using post it notes during Study 2, a TEC Expert Workshop. The field notes and written data provided by respondents were analysed following a thematic template approach [15] that was informed by the literature review and subsequent research questions.

### *D) Research Ethics*

Ethical approval for this research was obtained by the Ageing Research Centre (ARC) from the Research Ethics Committee (REC), University of Limerick.

## V. Study 1: PPI Workshop

The first study undertaken as part of this research was a PPI Workshop that addressed the perceived TEC needs as well as factors that can affect the acceptance, adoption and use of TEC by older people.

*1) Research Context*

The PPI Workshop was designed to elicit an understanding of perceived health needs of older people, potential contributions of TEC to support these needs, and the factors that may affect uptake and use of TEC by older people.

*2) Participants' Information*

20 respondents participated with the PPI Workshop. They were made up of older people living in the community, some of whom have health conditions, some who do not, and others including carers. They were members of a PPI panel organised by the Ageing Research Centre (ARC) at the University of Limerick. During the PPI Workshop they were organised into four groups of four to six people.

*3) Procedures*

The PPI Workshop participants were asked to consider TEC products and services that older people use or envisage that they may use in the future. They were also asked to consider factors that would be likely to affect their acceptance and use of TEC products and services.

*4) Findings*

The PPI Workshop findings identified the following themes:

- The potential of household technologies
- Safety and Security Issues
- Health Monitoring Tools
- Psycho-social considerations of TEC tools

*A) The potential of household technologies*

Initially, the PPI Workshop respondents considered everyday household devices such as video doorbells, which they reported to be potentially useful as an alert system. However, they also noted that doorbell cameras are ineffective if suspected perpetrators are not close enough to them. They also discussed technologies that can be used to control household devices such as switches. Some Study 1 respondents reported that voice controlled, and voice recognition of some household functions were desirable. However, others raised concerns about their use. For example, remote controlled lights and accessibility tools were posited to have the capability of improving quality of life. Temperature control in the home was also reported to be important for comfort and safety. Others felt that they wanted to retain control and direct use of household equipment rather than relying on voice activated technologies to control them. Similarly, Alexa, and similar voice-activated technologies were considered useful and helpful by some respondents, although there was inconsistency about its capabilities and functions. In fact, other PPI respondents felt that Alexa was not always reliable, particularly for more complex tasks such as scheduling appointments.

*B) Safety and Security Issues*

PPI respondents then addressed the use and potential of home safety and security. For home safety, they considered smoke alarms to be essential for safety. They reported that smoke and carbon monoxide detectors need to be functional and reliable. The discussion of Study 1 respondents on home security focused on security cameras and the use of online tools. They felt that security cameras alone were not sufficient to deter incidents. They also discussed security of online banking and risks of hackers and scams. In general, they expressed concerns about the security of personal data and access controls collected from any source. There was also strong resistance to the idea of constant surveillance and tracking of people through technologies.

*C) Health Monitoring Tools*

Wearable devices for health monitoring and assistance were also discussed by the PPI respondents in Study 1. Respondents raised potential issues about these with concerns about privacy and feeling under surveillance. Concerns about the increasing reliance on technology for essential services like healthcare were also expressed by respondents. Regardless of their age, some respondents felt that they were not old enough to need such technologies.

Pendant alarms were acknowledged as a common health monitoring tool, which informed views on features and functions of health monitoring technologies. The use and features of wearable health monitoring technologies repeatedly flagged a number of preferences including: that they should be affordable; they should be non-invasive wearables used exclusively for health monitoring and assistance; they should not include social features (e.g., communication, data sharing) nor require daily attention such as inputting information or requiring constant interaction. The PPI respondents noted that some smart watches provide regular alerts (e.g., fall risk) and some users would want to disable these as they can cause false alerts that can result in panic among users and carers. Additional issues raised about health monitoring technologies in Study 1 included concerns that an adverse health event could occur in unmonitored locations. The PPI respondents also expressed concern about a device that might be unable to directly contact a spouse or support worker in case of emergency.

*D) Psycho-social considerations of Enabl TEC tools*

Several psycho-social dimensions relating to the acceptance, adoption and use of TEC tools were posited by the PPI respondents in Study 1. These include perceived and actual competence in their use with the respondents reporting that competence levels in using technology can vary greatly. A lack of understanding on how to use technology could result in a fear of it. Therefore, there was a perceived need for better training and feedback in the use of TEC. These were also reported to relate to levels of confidence in the use of TEC. Study 1 also flagged the potential of fear of technology leading to social isolation. Even for competent and confident users of TEC, there was an expressed desire to maintain independence and not rely on technology for everything.

It was noted that older people often rely on younger individuals for tech support. In fact, PPI participants reported that technology use for people over 75 years old to be largely non-existent. Cognitive aspects were also considered in Study 1 including difficulties remembering passwords and navigating the fast pace of technological change. Related to this, it was reported that updates to devices often change the user experience, causing frustration. Psychomotor and sensory functions of older people were also raised with declines in motor skills and visibility/hearing issues as potential barriers to using TEC. One potential mitigation against this was a view that some TEC users prefer to talk to tech rather than type when using TEC tools.

## VI. Study 2: TEC Expert Workshop

The second study undertaken as part of this research was a TEC Expert Workshop that addressed the perceived TEC needs as well as factors that can affect the acceptance, adoption and use of TEC by older people.

*1) Research Context*

Study 2 took place in the premises of a company who are exploring the potential of TEC products in social housing.

*2) Participants' Information*

18 respondents participated in the TEC Expert Workshop. Participants were all employees of the company.

*3) Procedures*

To prepare TEC Expert Workshop participants for the research activities, a series of presentations were made that outlined the topics of interest and scope of the study. The presentations included findings from the literature outlined in the Introduction. The aim of this presentation was to focus the workshop activities on TEC relevant literature. Following this, Study 2 participants were invited to take part in the workshop. The workshop used Affinity Diagrams and consisted of interactive sessions addressing questions that are discussed as part of the findings presented below.

*4) Findings*

The findings of the TEC Expert Workshop identified the following themes:

- Perceived TEC needs of older adults
- Design considerations for TEC
- Health and Safety at Home
- TEC Concerns

### *A) Perceived TEC needs of older adults*

In Study 2, the TEC Expert Workshop respondents were asked to consider what needs older adults potentially have in the home. The ability to do things independently dominated as a recurrent theme. Managing one's own mobility, health and medicines management were the most frequent anticipated needs reported by the TEC Expert Group. The second most reported needs by the TEC Expert Workshop respondents were managing hygiene, daily schedules and communication. Additional needs identified in Study 2 included safety, being able to cook, and accessing and using things easily and comfortably.

### *B) Design considerations for TEC*

Design features that TEC Expert Workshop respondents felt would make older adults feel comfortable about using TEC in the home were focused on ease of use, interface design, and reliable performance. Ease of use that enabled a positive user interaction and experience were notable comments. It was suggested that simple and minimalistic design features for ease of use could include, visual aids, use of pictures, and colour coding with appropriate style and size. Clear feedback was also considered be helpful to build confidence in TEC use. Study 2 respondents reported that the technology should be ideally non-intrusive or at least minimally intrusive. They also reported a preference for TEC to be part of currently used technologies and be integrated into everyday activities. Considerations regarding reliability features included whether the technology should be on a mobile phone or not. However, help with any TEC should have readily accessible support by phone.

### *C) Health and Safety at Home*

When asked about potential safety issues in the home that would concern older adults, the TEC Expert Workshop participants reported health emergencies, accidents and resulting injuries, and security concerns as dominant issues. Health emergencies were most prevalent with falling considered to be the most common concern. Other accidents and injuries for example, house fires and burning oneself were also reported with examples of cooking or making hot drinks as possible risk factors. House and personal security issues such as burglaries and even fear when answering the door were also suggested. Being alone when experiencing any safety issue was an exacerbating aspect of each of the proposed concerns identified in Study 2.

### *D) TEC Concerns*

The TEC Expert Workshop participants were asked to provide what they considered were some of the concerns about using TEC that older adults were likely to have. Privacy and safety concerns were core concerns considered by the TEC Experts. In addition, usability, including ease of use was also reported as a key concern. Complexity, particularly difficulty understanding and navigating instructions was posited by the Study 2 respondents. A number of other factors presented by the workshop participants included: reliability of the TEC; product expense; resistance to accepting the need for TEC and embarrassment at having to use it, along with embarrassment at not understanding it; and a wish to not be a concern to family members and carers were also considered to be important issues that would affect the uptake and use of TEC by older people.

## VII. Discussion

The two studies outlined above have provided contributions to knowledge on perceived use of TEC applications for older adults. Following the TAM [3] and

HITAM [4] models, the findings have identified that there is potential for technology to support health and well-being for older adults if important factors are considered.

For Study 1, the Public and Patient Involvement (PPI) workshop respondents reported several very specific issues. These are summarised below. Following the summary of findings a discussion of four topics, the PU, PEOU of TEC [3] privacy concerns and preferences about the use of TEC.

For the PPI Workshop respondents in Study 1, several very specific issues were noted. When considering everyday use of technologies such as camera doorbells, household device control (e.g., lights) and heating technologies that offered improved quality of life, convenience, reliability and comfort were notable preferences. However, regardless of the nature of the device, Study 1 respondents were clear that they wanted to retain control over the use of their households [10]. For home security and safety, reliable functioning (e.g., smoke alarms), quality of features such as security cameras, perceived risks of hacking (e.g., online banking), and passive surveillance stood out as concerns [5]. Health monitoring technologies raised several key issues. These included concerns for privacy and discomfort about wearable health technologies that could lead to constant surveillance. Other notable issues raised about wearable health technologies in Study 1 included cost concerns, potential invasiveness (i.e., privacy), and questions about their reliability if something should go wrong [12][13]. The PPI Workshop respondents in Study 1 were also adamant that health monitoring technologies should be easy to use and require low levels of maintenance and interaction (i.e., active use) [16]. They also considered that any data collected by TEC should be only what is necessary and there should be no social features as opposed to movement tracking devised for fitness. The PPI Workshop respondents in Study 1 also discussed issues about their perceived competence, related anxieties and their functional capacities to use TEC. They reported that competence in the use of TEC would be likely to affect the confidence of users, so adequate training and support was essential to ensure that people did not reject TEC due to anxiety or fear of its use [16]. Study 1 respondents also noted potential challenges of remembering passwords, adjusting to updated functions and displays, and motor (e.g., dexterity) and sensory (e.g., vision and hearing) limitations that could affect user experience of TEC. They offered potential mitigating features such as talk over type as ways of addressing some of the posited challenges.

One of the dominant aspects of the PPI Workshop in Study 1 was focused on factors affecting PU and PEOU in TEC. Many of the respondents recognised the potential benefits of technology for managing their health. These included an appreciate of tools such as Electrocardiography (ECG) monitors in wrist-watches which they were very happy about. PPI Workshop respondents expressed concern about the risk of falls [17] coupled with a lack of safety features in homes, particularly in bathrooms and on stairs. However, wearable devices for fall detection received mixed reviews. Some respondents found them helpful, but one respondent reported that they had abandoned them after negative experiences (e.g., false falling alert when walking fast or placing something down with force). Other respondents felt they did not need to wear one. Participants also expressed concerns about the potential for adverse medication incidents with polydrug use. They suggested that medication management technologies and tools for checking drug interactions, and flagging allergies upon prescription could be beneficial. The PPI panel also considered that apps or devices that promote mindfulness, encourage attentive walking, and help with minor medical care (e.g., treating bruises or cuts) would be useful. They also talked about having faith that Artificial Intelligence (AI) could solve the issue of prescriptions.

The PPI workshop discussion involved a mix of anticipation and apprehension about TEC among older adults. TEC was considered to be potentially helpful especially if it works in the background and doesn't intrude on daily life. In fact, the PPI workshop panel devoted a considerable amount of time to discussing seemingly pervasive concerns among the older adults regarding the intrusiveness of technology [10]. These concerns centred around the uncertainty of how collected information could be used or where it might go. However, the PPI participants emphasised that with a proper introduction to technology to help understand its benefits, affordability, and the availability of continuous training, help prompts, or customer support, barriers to entry could be reduced. The panel talked about the importance of educating older adults on using technology effectively, suggesting that clear explanations and support on how to use TEC could enhance adoption rates [16]. They also stressed the importance of maintaining an open mind towards new technologies, as the upcoming generations are expected to be more tech-savvy. From this it can be submitted that the focus of TEC products and services should be on user-friendly technology that addresses safety concerns without sacrificing privacy or independence [5][10].

The main home TEC needs of older adults identified by the TEC Expert respondents in Study 2 focused on quality of life issues [11] that would enable continued independence while managing mobility, health and medicines and remaining safe at home. Independence focused on Activities of Daily Living (ADLs) such as hygiene, personal care, nutrition and managing daily schedules and appointments. Design feature priorities of TEC were reported by the Expert respondents as requiring ease of use to enable positive interactions with TEC. Confidence building features were also considered to be important to encourage TEC use. It was also reported that TEC should be non-intrusive, reliable with support available, and that the functions and features of TEC should match the expectations of normal everyday activities. These design features could enable enhanced PU and PEOU among older people. Beyond design features, the TEC Experts reported on health and safety at home with health emergencies, falling and other accidents and injuries, being of key concern. Home safety and particularly fire risks were also highlighted. Similarly, home security such as burglaries were noted. Concerns about being alone should a health, safety or security incident occur were highlighted too. In Study 2, the discussion continued to examine TEC concerns. These focused on privacy, complexity and ease of use, reliability of technology, costs associated with TEC products and services [5]-[13]. Resistance to accepting the need for TEC, embarrassment around its use, and not wanting to concern carers were also commonly reported.

The TEC Expert Workshop respondents did not offer the same level of detail as the PPI Workshop panel. However, the comments and responses largely echoed the issues raised in each workshop and aligned with TAM [3] and HITAM [4]. In fact, for both groups the principal factors that could affect the acceptance, adoption and continued use of TEC were privacy concerns, cost of products and services, and ease of use [5]. These key issues and others raised by the PPI panel and TEC Experts will inform the next phase of our research.

While the findings above demonstrate progression in knowledge and understanding of TEC by user groups and expert developers following TAM [3] and HITAM [4] models, it is worthy to note methodological features of this research that may have affected our findings. It is evident that the TEC Expert Workshop participants were briefed on the findings from the literature review. Naturally, this approach could prime and bias TEC Experts' responses. However, this is not considered as problematic for the current research. Progress towards practical outcomes in action research was facilitated by continually advancing the discussion within and between research participants. This enabled knowledge to be developed and refined in a structured way. The dialogical nature of each of the workshops supported the active and genuine participation of the PPI and TEC Expert groups. The workshops were collaborative and facilitated the agency of participants who provided new knowledge [14].

## VIII. Conclusions and Future Work

The research set out to identify factors important to the acceptance, adoption and use of TEC for older people. Two workshops, a PPI Workshop and a TEC Expert Workshop, were undertaken to identify relevant factors. The findings were in line with the literature supporting the view that TAM and HITAM are dependent on PU and PEOU. Additional issues such as privacy, cost and independence were noted. The next phase of this research will use the findings reported here to advance knowledge through additional Human Centred Design workshops. These will introduce actual TEC products and services that will enable workshop participants to critically appraise PU, PEOU, and identify enablers and barriers to TEC acceptance, adoption and use.

## Acknowledgments

This research has been co-funded by Lero, the Science Foundation Ireland Research Centre for Software, and Ei Electronics, Shannon, Ireland. We are grateful for the participation of the PPI panel and TEC Expert groups.

## References

[1] Global strategy and action plan on ageing and health. Geneva: World Health Organization; 2017. Licence: CC BY-NC-SA 3.0 IGO.

[2] S. T. MacMahon and I. Richardson, "Pathways, technology and the patient-connected health through the lifecycle," Front Digit Health, vol. 19, no. 5, Oct. 2023.

[3] F. D. Davis, "Perceived usefulness, perceived ease of use, and user acceptance of information technology," MIS Quarterly, vol. 13 no. 3, pp. 319-340, 1989.

[4] A. A. AlQudah, M. Al-Emran, and K. Shaalan, "Technology acceptance in healthcare: A systematic review," Applied Sciences, vol. 11, no. 22, 2021.

[5] O. Zafrani, G. Nimrod and Y. Edan, "Between fear and trust: Factors influencing older adults' evaluation of socially assistive robots," International Journal of Human-Computer Studies, vol. 171, pp. 1-14, 2022.

[6] G. Dermody, R. Fritz, C. Glass, M. Dunham, and L. Whitehead, "Family caregiver readiness to adopt smart home technology to monitor care-dependent older adults: A qualitative exploratory study," Journal of Advanced Nursing, vol. 80, no. 2, pp. 628-643, 2023.

[7] H. N. Kim, "Characteristics of technology adoption by older adults with visual disabilities," International Journal of Human-Computer Interaction, vol. 13, no. 37, pp. 1256-1268, 2021.

[8] D. Liu, A. Liu, and W. Tu, "The acceptance behavior of new media entertainment among older adults: Living arrangement as a mediator," The International Journal of Aging and Human Development, vol. 91, no. 3, pp. 274-298, 2020.

[9] T. Ohashi, M. Watanabe, Y. Takenaka, and M. Saijo, "Real-time assessment of causal attribution shift and stay between two successive tests of movement aids," Integrative Psychological and Behavioral Science, vol. 55, pp. 541-565, 2021.

[10] K. M. Sozio, "Exploring Older Adults' Perceptions of Issues Influencing Their Decisions to Adopt Gerontechnology," University of Arizona Global Campus ProQuest Dissertations and Theses, 2022.

[11] K. L. Rush et al., "Telehealth Use for Enhancing the Health of Rural Older Adults: A Systematic Mixed Studies Review," Gerontologist, vol. 62, no. 10, pp. 564-577, 2022.

[12] K. Kabacińska et al., "Functioning better is doing better: older adults' priorities for the evaluation of assistive technology," Assistive Technology, vol. 35, no. 4, pp. 367-373 2023.

[13] A. Franke, E. Nass, A. K. Piereth, A. Zettl, and C. Heidl, "Implementation of Assistive Technologies and Robotics in Long-Term Care Facilities: A Three-Stage Assessment Based on Acceptance, Ethics, and Emotions,". Front Psychol, vol. 26, no. 12, 2021.

[14] R. Ørngreen and K. T. Levinsen, "Workshops as a Research Methodology," Electronic Journal of E- Learning, vol. 15, no. 1, pp. 70-81, 2017.

[15] V. Braun and V. Clarke, "Using thematic analysis in psychology," Qualitative Research in Psychology, vol. 3, no. 2, pp. 77–101, 2006.

[16] J. Alpert et al., " Identifying barriers and facilitators for using a smartwatch to monitor health among older adults," Educational Gerontology, vol. 50, no. 4, pp. 282-295, 2024.

[17] G. Tónay, T. Pilissy, A. Tóth, and G. Fazekas, "Methods to assess the effectiveness and acceptance of information and communication technology-based assistive technology for older adults: a scoping review," International Journal of Rehabilitation Research, vol. 46, no. 2, pp. 113-125, 2023.

# Qawqaa: Aural Rehabilitation System for Children with Cochlear Implant Using Virtual Reality (VR)

Leenah AlMashari
*Department of Software Engineering*
*King Saud University*
Riyadh, Saudi Arabia
442200896@student.ksu.edu.sa

Sadeem AlSudais
*Department of Software Engineering*
*King Saud University*
Riyadh, Saudi Arabia
salsudais@ksu.edu.sa

Meriam Kefi
*Department of Software Engineering*
*King Saud University*
Riyadh, Saudi Arabia
mkefi@ksu.edu.sa

*Abstract*—**Qawqaa is a virtual reality based game for accessible and engaging aural rehabilitation in Arabic for children from eight to twelve years old with bilateral cochlear implants. The system contains two parts providing the game side that aims to enhance auditory skills, language development, and overall rehabilitation experience, and a monitoring and tracking side for rehabilitation center therapists, facilitating efficient tracking of a child's progress within the game. Using Qawqaa, we minimize clinic visits, enable effective progress monitoring, and ensure an enjoyable rehabilitation experience for the children.**

***Keywords-virtual reality; healthcare; cochlear implant; gamification; digitalization.***

## I. INTRODUCTION

The use of modern technologies across various domains is widely acknowledged for its ability to enhance service quality. Saudi Arabia's 2030 vision is currently spearheading digitalization efforts across industries, notably in healthcare, through comprehensive transformation programs. This initiative prioritizes enhancing care quality, improving patient experiences, and promoting sustainable health development at an international level [1].

In recent years, Virtual Reality (VR) has emerged as a valuable tool in numerous medical fields. VR is the use of computer technology to create simulated environments, immersing users in three-dimensional experiences instead of simply viewing a screen. The use of VR in medicine has shown success in various clinical applications, such as rehabilitation programs that help individuals recover and regain motor skills after injuries or neurological conditions. Rehabilitation services have been rolled back or stopped in most hospitals during COVID-19 [2]. It has had a significant impact on patients with disabilities emphasizing the need for digital home-training kits tailored to their requirements [3].

For people with Cochlear Implants (CI), rehabilitation after cochlear implant surgery is crucial to maximize its benefits. A cochlear implant is a device surgically implanted in the inner ear to stimulate the hearing nerve directly. It is suitable for individuals with severe to profound hearing loss in one or both ears, for whom hearing aids no longer provide sufficient benefit [4].

Children born with severe to profound sensorineural hearing loss, which occurs in approximately 1.5 per 1,000 cases globally [5], experience significant challenges in speech and language development. Aural rehabilitation helps children identify sounds and their meanings, while speech therapy aids in developing and understanding spoken language [6].

Gamification is the process of using game design elements and principles in non-game settings to captivate and motivate individuals, enhance desired behaviors, and improve user experiences. By integrating these game-like elements, rehabilitation programs become more interactive and enjoyable for patients, increasing their adherence and overall involvement in the treatment process [7].

Some existing research has used VR, mobile applications, or websites for CI rehabilitation. However, these systems lack monitoring and assessment features for specialists. For instance, Bears is a new VR game designed to enhance spatial hearing and speech perception in noisy environments, but it is not a fully integrated system with a monitoring component [8]. Ranan is an Arabic web-based tool used for clinical training, but it lacks the immersive experience of a 3D environment and is only usable in a clinic under specialist supervision [9]. Lastly, Karawan, a mobile app, offers multiple skills training but does not provide progress monitoring for the child [10].

So, we propose our solution, Qawqaa, a VR-based hearing training system in Arabic for children with bilateral cochlear implants. Qawqaa is designed as a home-based training solution that helps children overcome challenges related to (C1) limited access to rehabilitation services and (C2) provides the necessary training in an engaging, interactive environment filled with sounds, words, and phrases. Children advance through different stages of the game, leveling up as they progress. Unlike existing solutions, Qawqaa includes a dedicated therapist module, filling a critical gap by enabling therapists to monitor each child's progress through detailed scores and performance reports, making oversight easy and effective. The game is designed to improve the child's hearing and language reception skills, including sound localization, speech discrimination, and recognition. Overall, Qawqaa aims to enhance the child's abilities, reduce the need for clinic visits, and make the training process an enjoyable experience.

The rest of the paper is structured as follows. In Section II, we present the system overview. Section III shows the demonstration. We conclude the work in Section IV.

(a) Sound localization game (b) Word discrimination game (c) Speech recognition game

Figure 1. Three auditory training games

## II. SYSTEM OVERVIEW

In this section, we discuss the technical details of the Qawqaa system. Qawqaa has two main pieces of software, a VR game that serves as a training game for children, and a website used by specialists for monitoring children's progress in the games. These two parts are connected by a database. In this paper, we will go through the VR game only.

The Qawqaa VR game provides an interactive environment for children to stimulate rehabilitation training in an immersive and enjoyable way. The game has three skill modules to train as listed below.

Sound Localization: The ability to correctly localize sounds is an important feature of the auditory system directly linked to the ability to extract binaural information from the sound. In such situations, sound localization can help a listener quickly identify and orient themselves toward the talker in a group conversation. This is particularly important for CI users, because other cues for speaker identity, such as voice pitch, are diminished [11]. Here, we train the child's ability to hear a sound and localize the source of that sound with noise in the background in a 3D maze park, as shown in Figure 1(a), from where the child should get out by following the sound as an engaging element. The game gets harder with a louder noise in the background and more choices for sound sources.

Word Discrimination: Word Identification is the ability to accurately and automatically identify sight words and apply decoding strategies to read unfamiliar words. Auditory or speech perception focuses on auditory perception and sensory integration. It is composed of musical exercises that help one to identify and discriminate between different components of sound, such as time, tempo, duration, pitch, rhythmic patterns, and speech [12]. Here, we train the child with several similar words in sound, then he should choose the meaning as a picture of this sound. Figure 1(b) shows the sound icon with three images to choose from. This training combines the ability to differentiate between similar sounds and understanding the meaning of the sound. The difficulty in this skill is based on increasing the number of choices and the similarity between words' pronunciation. For example, both level one and two have two choices for each question, but the questions in level one consists of one word where level two consists of two words like the red flower and yellow flower. For level three, there are three choices with three words in the question such as the circled red painting, yellow squared painting, and green rectangular painting.

Speech Recognition: This represents the ability of an individual with hearing loss to accurately perceive and understand spoken language using technological aids or therapeutic interventions. It involves the process of converting spoken words or sounds into text or visual representations, allowing individuals with hearing impairment to access auditory information more effectively [13]. In this training game, we simulate the farming experience, as depicted in Figure 1(c), through three levels which are danging, seeding, and collecting crops. The child faces different words related to farming through voice commands given to him to assess his ability to recognize words and then follow commands. The difficulty for the three levels are increased by adding more words to sentences. For example, in level one a command given is (put the seed). For level two, the command is longer (the bean is good for the environment, put it in the seed). Lastly, in level three, a command is (the yellow corn on the top of the plant, put it in its box).

## III. DEMONSTRATION

In this section, we show how a child interacts with the system using a sound localization scenario.

In a 3D environment, we simulate spatial hearing to enable the child to detect the source of sound in a maze theme park, as shown in Figure 1(a), which consists of three levels, beginner, intermediate, and advanced. The difficulty level is based on the background noise level, where the beginner level has have little noise, the intermediate level has moderate noise, and the advanced level is very noisy. For engagement purpose, when the child detects the source of sound using the controller, he scores points. The assessment is based on the ability to correctly define the sources. After completing five questions in the level, he will get out of the maze and a reward board will be shown.

## IV. CONCLUSION AND FUTURE WORK

In this paper, we present Qawqaa as a step toward advancing aural rehabilitation for Arabic-speaking children. By offering a home-based training solution, Qawqaa reduces the need for frequent clinic visits while allowing therapists to efficiently monitor each child's progress. Looking ahead, we plan to expand Qawqaa by integrating artificial intelligence for advanced data analysis, providing children with more detailed insights into their progress. Additionally, we aim to introduce an interactive environment where children can engage with each other through a virtual world and leaderboard system, making the rehabilitation process more engaging and rewarding.

## ACKNOWLEDGMENT

We would like to thank Malath AlSaif, Nouf AlMuammer, Shahad AlJandal, Hanin AlHoti, and Tuwaiq Academy for their support.

## REFERENCES

[1] "Health Sector Transformation Program," Vision 2030, [Online]. Available: https://www.vision2030.gov.sa/en/vision-2030/vrp/health-sector-transformation-program/. [Accessed: 24-Sep-2023].
[2] R. H. Al-Khalil, "The impact of the COVID-19 pandemic on rehabilitation services provided for cochlear implant recipients in Saudi Arabia," *Annals of Medicine*, vol. 55, no. 1, pp. 869-880, 2023.
[3] A. Ilyas, A. Naiz, T. Abualait, and S. Bashir, "The Impact of COVID-19 Pandemic on Rehabilitation Services in a Tertiary Care Hospital in the Eastern Region of Saudi Arabia: A Single-Center Study," *Cureus*, vol. 13, no. 9, e18303, Sep. 2021. doi: 10.7759/cureus.18303.
[4] M. C. Staff, "Cochlear implants," *Mayo Clinic*, 10 May 2022. [Online]. Available: https://www.mayoclinic.org/tests-procedures/cochlear-implants/about/pac-20385021. [Accessed: 24-Sep-2023].
[5] G. Choe, S. K. Park, and B. J. Kim, "Hearing loss in neonates and infants," *Clinical and Experimental Pediatrics*, vol. 66, no. 9, pp. 369–376, 2023. doi: 10.3345/cep.2022.01011.
[6] F. Bergeron, A. Berland, D. Demers, and S. Gobeil, "Contemporary Speech and Oral Language Care for Deaf and Hard-of-Hearing Children Using Hearing Devices," *Journal of Clinical Medicine*, vol. 9, no. 2, p. 378, 2020.
[7] J. Janssen, "Gamification in Physical Therapy: More Than Using Games," *Pediatric Physical Therapy*, vol. 29, no. 1, pp. 95-99, 2017.
[8] L. Picinali, "Designing the BEARS (Both Ears) virtual reality training suite for improving spatial hearing abilities in teenage bilateral cochlear implantees," *Computational Audiology Network*, [Online]. Available: https://computationalaudiology.com/designing-the-bears-both-ears-virtual-reality-training-suite-for-improving-spatial-hearing-abilities-in-teenage-bilateral-cochlear-implantees/. [Accessed: 10-Oct-2023].
[9] A. Hagr et al., "The effect of the Arabic computer rehabilitation program 'Rannan' on sound detection and discrimination in children with cochlear implants," *Journal of the American Academy of Audiology*, vol. 27, no. 5, pp. 380-387, 2016.
[10] "Arab Karawan," Arab Karawan, [Online]. Available: https://arabkarawan.com/. [Accessed 20-Nov-2023].
[11] B. A. Wright and Y. Zhang, "A review of the generalization of auditory learning," *Philosophical Transactions of the Royal Society of London, Series B: Biological Sciences*, vol. 364, no. 1515, pp. 301–311, 2009.
[12] S. Kerber and B. U. Seeber, "Sound localization in noise by normal-hearing listeners and cochlear implant users," *Ear and Hearing*, vol. 33, no. 4, pp. 445–457, 2012. doi: 10.1097/AUD.0b013e318257607b.
[13] S. L. Bysak, R. S. Tyler, and S. L. Slevelmith, "Rehabilitation of speech recognition," in R. R. Fay and A. N. Popper, Eds., *Speech Processing in the Auditory System*, pp. 271-307, Springer, 2010.

# Fundamental Skills for Learning Strategies Based on Multiple Intelligences

Based on 192 Cases of Problematic Behavior in Vocational Training

Norikatsu Fujita
Ability Development Dept
Polytecnic University of Japan
Tokyo, Japan
e-mail: fujita@uitec.ac.jp

Sho Aoki
Electronics and Information Systems Dept
Shikoku Polytecnic College
Kagawa, Japan
e-mail: s.aoki0225@gmail.com

*Abstract*— **The growing number of trainees with developmental disabilities and other special needs in vocational training programs necessitates innovative approaches to support their learning and development. Despite individualized measures, many trainees with special needs struggle to achieve independent task performance. This study proposes a novel framework of fundamental skills for Learning Strategies, based on Multiple Intelligences (MI) theory, to address this challenge. By identifying and nurturing trainees' unique strengths from a set of skills that facilitate Learning Strategies, this framework aims to help them develop effective self-regulated learning strategies. The ultimate goal is to equip trainees with special needs to become independent learners, capable of completing tasks without relying on the support of instructors.**

***Keywords** - Multiple Intelligences(MI) Theory; Learning Strategies; Polytechnic science; Steps for Coding and Theorization (SCAT); Developmental disabilities.*

## I. INTRODUCTION

Effectively educating individuals with developmental disabilities and other special needs in Japanese vocational training schools is a pressing challenge. Traditional approaches often rely on analyzing individual cases and identifying causal relationships between specific factors and success or failure. However, these methods are limited in providing a comprehensive understanding of the diverse range of factors influencing educational outcomes.

In contrast, other fields have developed parameters to comprehensively grasp the characteristics of individuals and reflect them in national policies and education. While parameters for diagnosing developmental disabilities exist [2], there is a complete lack of parameters specifically designed from an educational perspective for vocational training.

Given the pressing challenge of effectively educating individuals with developmental disabilities in vocational training schools, this study aims to develop a comprehensive framework of fundamental skills for Learning Strategies, specifically designed for vocational training settings, to address the educational needs of these individuals.

In a previous study [1], we identified 32 fundamental skills for Learning Strategies that were associated with 192 cases of problematic behaviors hindering training activities in vocational training. Fundamental skills are essential skills that facilitate the development of effective Learning Strategies. To refine these skills and enhance their applicability to vocational training settings, we conducted a SCAT (Steps for Coding and Theorization) analysis on interview data from 11 vocational training instructors. This analysis allowed us to merge, split, delete, rename, and add skills, resulting in a revised set of 26 fundamental skills. The revised set of 26 fundamental skills offers a practical and effective framework for vocational training instructors.

The rest of the paper is structured as follows. In Section II, we present the methodology used to refine the Fundamental Skills, including the SCAT analysis and the integration of Multiple Intelligences theory. In Section III, we describe the results of the refinement process, highlighting the specific changes made to the skill set. Finally, we conclude the work in Section V, summarizing the key findings and discussing future directions for research and implementation.

## II. METHOD

We conducted one-hour interviews with 11 vocational training instructors (hereafter referred to as instructors) who had at least two years of experience in the field. The purpose of these interviews was to clarify and refine the range of 32 Fundamental Skills that were associated with problematic behavior in 192 cases [3].

Instructors were asked to imagine individuals with significantly low or high evaluations of these skills and to describe their behaviors in the vocational training setting, including during breaks.

SCAT is a qualitative data analysis method developed by Ohtani [4]. It is used to analyze segmented qualitative data, such as linguistic data from interviews. The goal of SCAT is to construct a narrative by weaving together thematic constructs. The method consists of four coding and analysis steps, which are performed manually:

(1) Notable Words and Phrases:

Identify keywords and phrases that are significant and related to the Fundamental Skills. These are the specific terms and expressions that stand out in the interview data.

(2) Paraphrasing:

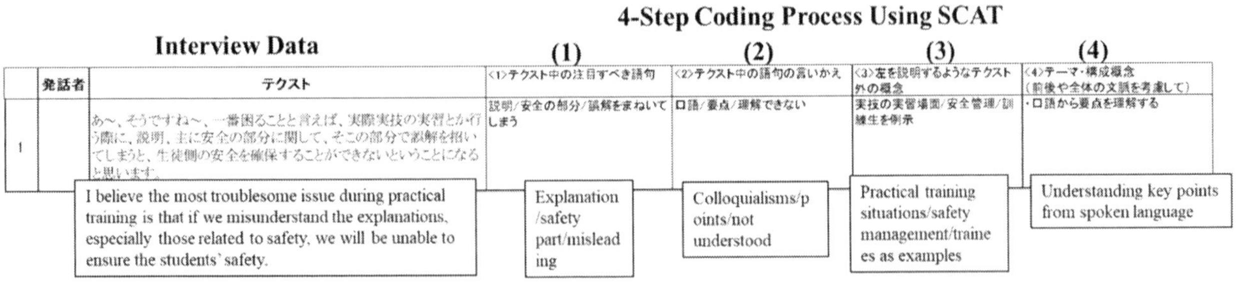


Figure 1. Concepts of the results of SCAT of the 11 interviews about "Understanding key points from spoken and written language".

**Understanding key points from spoken and written language**

①Extracting main points from conversations and texts
②Understanding the main points from oral language
③Understanding explanations
④Facilitating understanding of the purpose from colloquial language Extracting
⑤important words and phrases from conversations and sentences
⑥Does not understand the main points and subjects of conversations
⑦Cannot extract main points from sentences
⑧Cannot extract main points from spoken words and sentences
⑨Cannot understand the main points from colloquialisms
⑩Cannot understand the other person's intention in conversation
⑪Misunderstands main points from colloquialisms
⑫Cannot understand the main points from sentences
⑬Inability to extract important words and phrases from sentences

Figure 2. SCAT Results for "Understanding key points from spoken and written language".

Generalize these identified words and phrases. This step involves rephrasing the notable words and phrases to capture their broader meaning and context.

(3) External Concepts:

Explain the text using concepts from outside the interview data. This step involves bringing in relevant theories, ideas, or concepts from external sources to provide a deeper understanding of the interview data.

(4) Themes/Constitutive Concepts:

Develop themes or overarching concepts based on the previous steps. This final step involves synthesizing the paraphrased data and external concepts to create broader themes or constitutive concepts that capture the essence of the interview data.

For instance, in the four-step coding process, the notable words and phrases "Explanation", "Part of safety" and "Misleading" were identified in relation to the fundamental skill of "Extracting main points from conversations and sentences." These were then paraphrased as "Colloquial" and contextualized with practical training situations, such as safety management and examples of trainees. The overarching theme or constitutive concept was determined based on these elements and is labeled "Understanding the main points from colloquial."

Figure 1 provides an illustrative example of SCAT analysis, demonstrating how notable words and phrases are identified, paraphrased, and contextualized to develop overarching themes or constitutive concepts. The SCAT analysis of the Fundamental Skills aimed to establish a well-structured and coherent framework without redundancies. In consolidating the 32 Fundamental Skills into 26, the skill set was restructured to address three primary objectives: eliminating redundancies, clarifying ambiguous definitions, and incorporating essential new skills. By integrating overlapping skills, redefining ambiguous ones, and adding necessary elements, we created a more practical and applicable skill set. Furthermore, low-priority skills were either integrated or removed to enhance the overall efficiency and sentences.

Figure 2 shows the constructs from the SCAT analysis of the 11 interviews on "extracting key points from conversations and sentences. 13 different constructs were identified, indicating subtle differences in perception compared to the original Fundamental skills. There are individual differences in people's ability to extract key points from conversations and texts. Specifically, differences can be seen in how people identify information, their understanding of context, their ability to summarize, and many other aspects. Thus, each person processes information differently.

Figure 3 shows the process of redefining Fundamental Skills by clarifying the scope. This process entails creating, deleting, and integrating new Fundamental Skills to equalize the size of the scope as much as possible, and making changes to those that cover all problem behavior situations.

Through iterative refinement using the SCAT method, an initial set of 32 Fundamental skills was reduced to 26. This involved 13 skill name changes, 2 divisions, 6 mergers, 4 deletions, 12 modifications to skill descriptions, and 1 category shift, along with the addition of two new skills. The skill "Tacit understanding," initially categorized under interpersonal intelligence, was removed prior to data collection due to its reliance on socially acquired knowledge (common sense) that is difficult to quantify.

For instance, the skills "Distinguishing between objects" and "Distinguish between parallel lines and single lines," both under visual-spatial intelligence, were merged into " Identifying 3D shapes" and " Identifying 2D shapes," respectively. While the former originally assessed the ability to navigate physical spaces, instructor feedback primarily focused on the ability to visualize shapes and structures. Similarly, "Distinguishing between object " was initially intended to assess the ability to distinguish between objects like positive and negative drivers, but instructor responses frequently referred to the identification of line types in diagrams. Given the importance of this skill in vocational training, the two skills were combined, and the terms were adjusted to reflect the different dimensions involved. The skill "correct interpretation" under logical-mathematical intelligence was divided into "Subdivision of information" and " Relating to similar experiences." The former assesses the ability to break down information into smaller components, while the latter involves linking these components to prior knowledge. Skills such as "Image of completion" and "Grasping risky behavior" under intrapersonal intelligence were merged due to their predictive nature. However, "Image of completion," which involves visualizing a completed assembly from a diagram, was also related to visual-spatial skills. Furthermore, the concept "Understanding of jokes (playful language)" was added under linguistic intelligence to account for situations where trainees may misinterpret literal meanings in social interactions.

## III. RESULTS

Figure 4 compares the original and redefined Fundamental skills using SCAT. The original 32 skills were reduced to 26 by merging, splitting, deleting, renaming, and adding. Only the skills in bold remained unchanged: the 26 Fundamental Skills fall into the six categories of MI theory, with the exception of Musical Intelligence and Erudite Intelligence, which are less relevant to manufacturing. Each category contains four to five Fundamental Skills.

SCAT eliminates duplication and ambiguity in the Fundamental Skills and creates practical common skills. Fundamental Skills allow trainees to train using skills in which they are proficient. Instructors can also train trainees to match their skill characteristics. Addressing the challenges of dealing with struggling students requires more than individual instructor efforts. Many schools rely on consultations with experienced instructors or establish leader-centered support systems, particularly in schools for individuals with disabilities.

The Fundamental Skills align with Gardner's MI theory [5], which emphasizes individual strengths and learning styles. In the U.S., many educational practices utilize MI theory to help learners identify their strengths and address weaknesses. This approach is particularly beneficial for individuals with developmental disabilities who often exhibit uneven skill characteristics. Low self-esteem can hinder learning by limiting individuals' expectations and motivation. Developmental disabilities can contribute to low self-esteem due to peer undervaluation and self-doubt. Additionally, many individuals with developmental disabilities have co-occurring mental health conditions.

Educational practices for individuals with developmental disabilities have evolved, with emphasis on self-regulated learning and leveraging individual strengths. Dennis Laird and others have advocated for teaching individuals with developmental disabilities to capitalize on their preferences [6].

The educational practice of MI theory emphasizes establishing personalized learning approaches that leverage individual strengths. When necessary, weaknesses should be complemented with strengths. Many trainees with developmental disabilities or other special needs struggle with uneven skill characteristics. The goal is to address these challenges by capitalizing on their strengths.

## IV. DISCUSSION

In this paper, we defined fundamental skills based on 192 examples of problematic behaviors observed in actual workplace training situations. Unlike traditional key competencies and basic skills for working adults, which are based on expert opinions, we defined fundamental skills from multiple perspectives through qualitative analysis of actual problematic behaviors. In particular, the interview-based analysis provided an understanding of the diversity and depth of the skills, ensuring their reliability. Furthermore, classifying the extracted fundamental skills according to the Multiple Intelligences (MI) theory confirmed that the various intelligences in the MI theory effectively explain the diversity of the skill groups obtained in this study. The skill system was structured similarly to traditional basic work skills and key competencies, with each category containing multiple elements.

The issue addressed in this study is that students with special needs are unable to complete tasks without teacher

| MI | Fundamental Skills | Scope | Reason for improvement | Refined Skills name | supplementary explanation |
|---|---|---|---|---|---|
| Linguistic-verbal | Extracting key points from conversations and texts | · Smoothly understand the purpose from colloquial language.<br>· Inability to understand the other person's intention in conversation. | The term "oral" was changed to "spoken" because many situations were answered in which the instructor gave one-sided explanations rather than conversing. | Understanding key points from spoken and written language | Insert meaning of how correctly grasp the purpose from oral explanations and written questions. |
| | Understanding instructions | · Cannot understand the object of "this" and "that".<br>· Can understand the object of "this" and "that". | | Comprehension of directives (maintenace of skill name) | Insert meaning of how grasp of the subject of "this" and "that". |
| | Comprehension of abstract expressions | · Cannot generalize and understand mathematical properties.<br>· Understand abstract expressions. | | Comprehension of abstract expressions (maintenance of skill name) | Insert meaning of how correctly interpret and use ambiguous expressions. |
| | Basic Kanji Proficiency | · Cannot comprehend Kanji characters at the vocational level.<br>· Can read and write daily life level Kanji characters. | The answer was changed to "Common Kanji(Jyoyo Kanji)" because many answers indicated the vocational training level and at the same time indicated that the vocational training level was within the scope of study in compulsory education. | Reading and writing of kanji for common use | Insert meaning of how to read and write kanji used in daily life. |
| | | | Added skills: Because there are situations where I have trouble getting a joke across in conversation with people. | Understanding of jokes (playful language) | Insert meaning of how understand the humor of others. |
| Logical-mathematical | Sequential action and thinking | · Can work according to the established sequence.<br>· Deviates from work procedures. | Transition to "Subdivision of information" and "Understanding priorities". | | |
| | Grasp of priorities | · Can determine priorities based on efficiency.<br>· Cannot determine priorities based on deadlines. | Integration of "grasp of priorities" and "grasp of important parts": because the fundamental skills of both is "Grasp of priorities". | Grasp of priorities (integration of skills) | Insert meaning of how set the level of importance considering the conditions. |
| | Grasp of important parts | · Can grasp the important parts of things.<br>· Can understand the important parts of multiple tasks. | Integration of "grasp of priorities" and "grasp of the important parts": because the fundamental skills of both is "Grasp of priorities". | Grasp of priorities(integration of skills) | (Insert meaning of how set the level of importance considering the conditions.) |
| | Subdivision of information | · Cannot separate information and understand its components.<br>· Cannot separate work instructions and understand procedures. | | Subdivision of Information (maintenance of skill names) | Insert meaning of how analyze information to identify its components. |
| | Completion of omissions | · Can understand the lack of processing and improve it.<br>· Does not have the knowledge to accomplish the objectives. | Because many "Unknown points" were extracted as a composition concept. | Completion of unclear points | Insert meaning of how compensate for missing information by asking others, researching, etc. |
| | Basic mathematical skills | · Can perform four arithmetic operations including fractions.<br>· Can perform the four arithmetic operations. | The reason is that "Four arithmetic operations" was extracted as a component concept in many cases. | Four arithmetic operations | Insert meaning of how perform calculations of daily life. |
| | Correct interpretation (of logical and mathematics) | · Make sense of numbers by calculation.<br>· Interpret the significance of things correctly. | Transition from "(Logical-mathematical) correct interpretation" to "Subdivision of information" and "connection to similar experiences". | | |
| Visual-spatial | Recognition of three-dimensional arrangements | · Inability to perceive three-dimensional distance and depth.<br>· Cannot visualize three-dimensional shapes. | | 3D identification | Insert meaning of how correctly perceive location and space. |
| | Short-term memory of objects | · Cannot memorize numerical values in the short term.<br>· Can memorize the shape of objects in the short term. | Memory is difficult to handle not only objects but also a wide range of numerical values, words, and phrases. | | |
| | Identification of objects | · Understanding of line types is a prerequisite.<br>· Cannot distinguish shapes. | | 2D identification | Insert meaning of how Correctly recognizes shapes and line types of figures. |
| | Color recognition | · Cannot distinguish colors close to the primary color.<br>· Cannot distinguish colors. | The reason for this is that many of the answers were based on situations that require not only recognition but also identification, such as color codes and wiring. | Color identification (maintain skill names) | Insert meaning of how distinguish colors. |
| | Extraction of key points from | · Cannot focus on hazardous areas.<br>· Cannot focus on important points. | The skill was changed to evaluate the ability to grasp information intuitively from visual information, since it is a characteristic of developmental disabilities to focus on details and not be able to overview the whole picture. | Intuitive grasp of the big picture | Insert meaning of how recognize the big picture and then check the details. |
| Body-kinesthetic | Manual dexterity | · Handling of fine parts.<br>· Errors are introduced by precise work. | | Manual dexterity(maintain skill names) | Insert meaning of how handle small parts. |
| | Hand agility | · Manual dexterity is more important than manual agility.<br>· Perform detailed work smoothly. | | Manual agility(maintain skill names) | Insert meaning of how finish the work within the specified time. |
| | Body image | · Imitates the body movements of others.<br>· Does not have proper awareness of body range. | The body image is not a common word, so it was changed to an easy-to-understand word. | Correct recognition of body movements (body image) | Insert meaning of how awarenes of invisible areas. |
| | Coordinated actions | · Can work with awareness of multiple parts of the body.<br>· Can work with awareness of multiple parts of the body. | | Coordinated actions(maintain skill names) | Insert meaning of how work with awareness of multiple locations. |
| Interpersonal | Face to face with others | · Inability to use appropriate body language.<br>· Inability to express appropriate emotions. | The concept was changed to a skill that communicates because it included a lot of conversation and body language. | Communication of accurate information | Insert meaning of how communicate facts accurately. |
| | Resistance to interpersonal interaction | · Is uncomfortable in conversation.<br>· Cannot establish positive rapport with new acquaintances. | Changed the content to emphasize the point of communicating without being overly nervous in front of others. | Normalcy in public | Insert meaning of how make eye contact with others. |
| | Understanding the other person's position and situation | · Cannot anticipate the feelings of others and act accordingly.<br>· Cannot understand the other person's situation. | "grasp of other person's position and situation" is divided into "grasp of other's intention" and "grasp of other's feelings". | Grasp of others' intentions | Insert meaning of how understand the atmosphere of a place. |
| | | | | Grasp of others' feelings | Insert meaning of how read other people's thoughts. |
| | Tacit understanding | | The reason is that it is difficult to formalize the knowledge. | | |
| | Caring for others | · Can build good relationships.<br>· Has difficulty in building good relationships. | Moved to "grasp of other's intentions" and "understanding of other's feelings". | Communication of emotions | Insert meaning of how express their emotions appropriately |
| Intrapersonal | | | Add skills: to be consistent with intuitive grasp of visual information. | Grasp of prospects | Insert meaning of how accommodate sudden schedule changes |
| | Connection to experience | · Cannot work following the example.<br>· Repeats the same mistakes. | The reason is that we believe that the accumulation of related experiences, rather than completely different experiences, will enhance our ability to apply them. | Connection to similar experiences | Insert meaning of how use your past experience to solve problems |
| | Image of completion | · Clearly defines the target quality.<br>· Can grasp the process from the image of completion. | Moved to "intuitive grasp of prospects". | | |
| | Change of mind | · Can switch from break to work quickly.<br>· Cannot let a bad event drag on. | Because emotions not only need to be switched, but also need to be controlled in diverse ways, such as suppressing anger. | Emotional control | Insert meaning of how turn on/off the negative emotions |
| | Grasp of cause-and-effect relationships | · Cannot investigate the cause of work failure.<br>· Cannot understand the structure of equipment. | Moved to"Connect to Similar Experiences". | Connection to similar experiences | Insert meaning of how past experience can be used to solve problems |
| | Viewing from another angle | · Flexible in dealing with problems instead of sticking to existing methods.<br>· Cannot consider multiple means to solve a problem. | Logical-mathematical intelligence is the ability to analyze things logically, and not sticking to one way of thinking can be achieved through "emotional control" and "Grasp of prospects". | | |
| | Grasp of risky behavior | · Can predict and avoid hazards.<br>· Can anticipate danger. | Moved to "Attention switching". | | |
| | Asserting oneself | · Cannot express physical condition.<br>· Needs to improve the work environment. | Moved to "emotional transfer" and "information transfer | | |

Figure 3. Process of redefining the Fundamental Skills by clarified the scope.

1) Extracting key points from conversations and texts
**2) Understanding instructions**
**3) Understanding of abstract expressions**
4) Basic Kanji Proficiency
5) Sequential action and thinking
6) Understanding of priorities
7) Understanding of important parts
8) Subdivision of work
9) Completion of omissions
10) Correct interpretation
11) Basic mathematical skills
12) Recognition of three-dimensional arrangements
13) Short-term memory for objects
14) Distinguishing between objects
15) Distinguish between parallel lines and single lines
16) Extracting key points from visual information
**17) Hand dexterity**
18) Hand agility
**19) Correct recognition of body movements (Body image)**
**20) Coordinated actions**
21) Face to face with others
22) Resistance to other people
23) Understanding the other person's position and situation
24) Tacit understanding
25) Caring for others
26) Connection to experience
27) Image of completion
28) Change of mind
29) Grasping cause-and-effect relationships
30) Viewing from another angle
31) Grasping risky behavior
32) Asserting oneself

**Original Fundamental Skills**

SCAT Analysis

| Multiple Intelligences | | Fundamental Skills | Examples of Skill Assessment Situations |
|---|---|---|---|
| Linguistic-verbal | 1 | Reading and writing Common Kanji | Can you read and write kanji used in daily life? |
| | 2 | Understanding key points from spoken and written language | Can the student correctly understand the purpose of oral explanations and written questions? |
| | 3 | **Understanding instructions** | Can the student correctly understand the object of "this" and "that"? |
| | 4 | **Understanding of abstract expressions** | Can you correctly interpret and use ambiguous expressions? |
| | 5 | Understanding of jokes (playful language) | Can you understand the humor of others? |
| Logical-mathematical | 6 | Basic arithmetic operations | Can you perform routine calculations? |
| | 7 | Subdivision of information | Can you analyze information and clarify its components? |
| | 8 | Completion of unclear points | Can you make up for missing information by asking others or researching? |
| | 9 | Understanding of priorities | Can you set the level of importance in consideration of conditions? |
| Visual-spatial | 10 | Identifying 2D shapes | Can you recognize shapes and line types of figures ? |
| | 11 | Identifying 3 D shapes | Can you recognize the position and space? |
| | 12 | Identifying color | Can the student distinguish colors? |
| | 13 | Intuitive understanding of visual information (holistically) | Can you recognize details after recognizing the whole image? |
| Body-kinesthetic | 14 | Hand dexterity | Can they handle small parts? |
| | 15 | **Hand agility** | Can you finish the work within the specified time? |
| | 16 | **Correct recognition of body movements (body image)** | Can you recognize invisible areas? |
| | 17 | **Coordinated actions** | Can you work with awareness of multiple locations? |
| Interpersonal | 18 | Maintaining composure in public | Can you make eye contact with others? |
| | 19 | Conveying accurate information | Can you communicate facts accurately? |
| | 20 | Expressing emotions | Can you express your feelings appropriately? |
| | 21 | Understanding of others' intentions | Can you read the thoughts of others? |
| | 22 | Understanding of others' feelings | Can you understand the atmosphere of a place? |
| Intrapersonal | 23 | Understanding future prospects | Can you handle sudden schedule changes? |
| | 24 | Relating to similar experiences | Can you apply past experience to solve problems? |
| | 25 | Emotional control | Can you switch on and off without letting negative feelings get in the way? |
| | 26 | Attention switching | Can you work without missing anything? |

**Redefined Fundamental Skills**

Figure 4. The original and Redefined Fundamental Skills redefined using SCAT.

support, even after repeated individualized interventions. To complete tasks independently, these students need to master strategies that leverage their own specialized skills.

Students need to plan, monitor, and evaluate their own learning process, making adjustments as needed. In training, students are first taught the importance of goal setting and are asked to set specific goals. Next, we teach them how to regularly monitor their progress and review their strategies for achieving these goals.

We also help them develop self-evaluation skills and enhance their sense of self-efficacy through successful experiences. Learners need to understand their own learning style and discover the best way to learn. fundamental skills serve as the link between learning strategies and their individual learning styles.

Furthermore, the fundamental skills are categorized by MI intelligence, allowing for the use of educational practices based on MI best practices. Even if one does not excel in a particular intelligence compared to others, it is important to master strategies that leverage the intelligence one feels confident in. In a survey of white, middle- and upper-income children, 76% had at least one strong intelligence, 30% had no weak intelligence, and 20% had no strong intelligence [7]. Even if one does not excel in a particular intelligence, it is important to master strategies using the intelligence one likes and feels confident in. People with developmental disabilities often have excellent intelligence in some areas while having weaknesses in others.

The inspiration for this study came from a case where a person with a lack of body awareness, who often fell and hit walls, became proficient in gymnastics and skating. This improvement was due to an encounter with an instructor who taught with logic. The instructor said, "Put your left foot on the line, and your right hand will come off naturally," or "You lost your balance because the tip of your nose was not facing the line." This approach emphasizes the importance of teaching based on strengths, where strengths complement weaknesses. Instructional training for trainees using fundamental skills in Polytechnic University has been conducted since 1998, highlighting the duplication and granularity of the 32 fundamental skills. We have been using this improved version since 2022 and have not encountered any major problems.

## V. CONCLUSION AND FUTURE WORK

This study has demonstrated the effectiveness of redefining fundamental skills using SCAT and aligning them with Gardner's MI theory to support trainees with special needs. By focusing on individual strengths and learning styles, we have created a more inclusive and practical approach to skill development. The improved version of the fundamental skills has been successfully implemented since 2022, with positive feedback from both trainees and instructors.

Future Work:

1) Longitudinal Studies: Conduct long-term studies to evaluate the sustained impact of the redefined fundamental skills on trainees' performance and independence in the workplace.

2) Broader Application: Explore the applicability of the redefined fundamental skills in different industries and

educational settings to ensure their versatility and effectiveness.
3) Technological Integration: Investigate the potential of integrating technology, such as AI and adaptive learning platforms, to further personalize and enhance the training experience for individuals with special needs.
4) Collaborative Efforts: Foster collaborations with other educational institutions and organizations to share best practices and continuously improve the fundamental skills framework.

By addressing these areas, we aim to further refine and expand the scope of our approach, ultimately contributing to a more inclusive and supportive learning environment for all individuals.

## ACKNOWLEDGMENT

This work was supported by JSPS KAKENHI Grant Number 22K02824.

## REFERENCES

[1] N. Fujita et al., "Developing a Computer-Based Vocational Training Environment that Complements the Weak Skills and Career Development of Trainees," International Journal on Advances in Intelligent Systems, pp. 279-289, 2018.

[2] P. D. Flanagan and A. S. Kaufman, "Essentials of WISC-IV Assessment," Hoboken, 2009.

[3] Japan Organization for Employment of the Elderly, Persons with Disabilities and Job Seekers, "Effective vocational training examples for people with developmental disabilities" Research Report No. 119, Japanese, 2007.

[4] Ohtani.SCAT Steps for Coding and Theorization: https://www.educa.nagoya-u.ac.jp/~otani/scat/ (in Japanese)

[5] H.Gardner, "Multiple Intelligences: The Theory in Practice," Basic Books; Reprint Edition edition, 1993.

[6] D. Reid and C. Green: Preference-Based Teaching: Helping People with Developmental Disabilities Enjoy Learning without Problem Behavior, Habilitative Management Consultants, Inc., 2005.

[7] H. Gardner, "Frames of Mind: The Theory of Multiple Intelligences," Basic Books, 2011.

# Mobile App Use among Chronic Pain Patients

A Gamified Approach

Anabela Marques
CHEDV - Centro Hospitalar de entre Douro e Vouga, E.P.E.
Santa Maria da Feira, Portugal
e-mail: anabela.marques@chedv.min-saude.pt

Duarte Duque
2Ai – School of Technology, IPCA
Barcelos, Portugal
LASI - Associate Laboratory of Intelligent Systems
Guimarães, Portugal
e-mail: dduque@ipca.pt

*Abstract*— **Chronic pain, which affects more than 30% of adults in Portugal, requires a multidisciplinary therapeutic approach. To stabilize or minimize the patient's pain, it is essential that patients comply with the prescribed treatments, not only pharmacological but also regarding physical activity. The assessment of the patient's condition, regarding the level of pain and their general physical condition, when carried out only in a consultation context, is not always reliably expressed by the patient. This study explores the potential of a mobile application in chronic pain management, with a focus on fibromyalgia. FisioQuest uses gamification to increase patient engagement and treatment adherence, including continuous symptom monitoring, physical exercise and cognitive-behavioral therapy. This application is based on preliminary studies that have shown significant improvements in pain severity, anxiety and depression among fibromyalgia patients who use mobile applications. This article details the requirements, specifications and use cases of the FisioQuest app, highlighting its potential benefits and future directions. It is hoped that, when validated, the FisioQuest app can offer an effective tool for improving patients' self-management and quality of life in the treatment of chronic pain.**

***Keywords - chronic pain; fibromyalgia syndrome; mobile app; gamification.***

## I. INTRODUCTION

Chronic primary pain is defined by the International Association for the Study of Pain as pain in one or more anatomical regions that [1] persists or recurs for more than 3 months and is associated with significant emotional distress (e.g., anxiety, anger, frustration, or depressed mood) or significant functional disability (interference in activities of daily life and participation in social roles), and the symptoms are not better accounted for by another diagnosis [2].

Chronic pain causes significant physical and emotional repercussions for the patient, as well as socio-economic consequences with extended periods of work absenteeism. Consequently, it has a huge negative effect on the lives of patients and their families, as well as having an important economic impact on society [13]-[16]. In Portugal, its prevalence exceeds 30% in the adult population [3].

The limited efficacy of pharmacotherapy in treating chronic pain and the long-term side effects of these pharmacological treatment options [2][4][5] motivate the search for new techniques and therapeutic approaches.

Among the non-pharmacological measures, which the guidelines emphasize should be the foundation of the therapeutic approach to chronic pain [3], are, among others, physical exercise, patient education about their condition, and psychotherapeutic intervention through Cognitive-Behavioral Therapies (CBT), the latter being cited as having a strong level of evidence by the American Society of Anesthesiologists [3].

Currently, technology influences all facets of modern society and is set to revolutionize healthcare in the future. The use of new technologies is growing in health communication for health promotion, disease prevention and health care delivery [6]. The COVID-19 pandemic has expedited the digitization of healthcare, fostering the development and implementation of telehealth methodologies, including remote medical consultations, home hospitalization, and remote monitoring of vital signs.

The global trend of increasing information and communication technology usage in healthcare enhances public health and well-being. These technologies have taken various forms, such as software applications, mobile apps, digital sensors, immersive virtual reality equipment, and other therapeutic interventions.

With the substantial rise in mobile device usage in recent years, mobile application-based treatment options have become increasingly prevalent. While the empirical validation of such app-based programs is still pending [4], they hold potential for empowering patients to independently manage their pain, thereby reducing the likelihood of relapse.

A recent systematic review and meta-analysis on mobile application-based interventions for chronic pain patients has been published [4]. The final sample comprised twenty-two studies, with a total of 4679 individuals. The results suggest that apps-based treatment can be helpful in reducing pain, especially in the long-term [4]. Thus, we understand that mobile applications can be a powerful tool to assist patients and healthcare professionals in the approach and treatment of Chronic Pain. This can be achieved through better perception of pain evolution throughout the day, the effect of medications and their side effects, the impact of the disease on daily living activities and the patient's professional life, sleep quality, and physical exercise performance. In addition to potentially improving the patient's quality of life, it allows for the optimization of communication and the relationship between doctor and patient, with evident benefits for all.

Fibromyalgia Syndrome (FMS) is a rheumatic disorder characterized by chronic widespread pain often associated with fatigue unrefreshed sleep and cognitive difficulties [6][7]. FMS is a common syndrome in the general population, reaching a prevalence of 2–3% worldwide [8]. It is a debilitating condition that impairs quality of life, increases health care utilization, impairs work productivity and daily activities [6].

Despite the acquired perception regarding the importance of mobile applications in assisting patients in managing their illness, little is known about persons with fibromyalgia use mobile apps for health-related purposes [6]. Demonstrating the relevance of the topic, in the last decade, we have observed the development of mobile applications specifically for patients with fibromyalgia, such as Fibroline [9] and Profibro [10]. Studies have been published analyzing the outcomes of using these applications in promoting self-care, monitoring symptoms, and improving quality of life [9][10].

A cross-sectional survey was recently published, analyzing the use of mobile apps among persons with fibromyalgia [11], concluding that approximately two-thirds of the sample used mobile apps. The results showed that ease of use and the fact that the apps were free were essential factors for their utilization.

On the other hand, reasons for discontinuing use included issues with privacy, the effort required, lack of interest, and lack of perceived quality [11]. Other reasons for app nonuse were lack of awareness and lack of knowledge on how to use them, indicating that disseminating information about apps and addressing other barriers, such as providing user support, are critical to increasing uptake [11].

Miro et al. [12] published a preliminary evaluation of a mobile-app-delivered, Cognitive Behavioral Treatment (CBT)-based intervention in helping adults self-manage fibromyalgia symptoms. The study included a total of 100 adults with FMS which were given access to the digital treatment program and downloaded the app. Pain severity, anxiety symptoms, depression symptoms, fatigue, and sleep quality were assessed. Data showed significant improvements in pain severity ($p = 0.007$, $d = 0.43$), anxiety ($p = 0.011$, $d = 0.40$) and depressive symptoms ($p = 0.001$, $d = 0.50$) from pre-treatment to post-treatment.

This paper aims to further explore the potential of mobile applications in the treatment of chronic fibromyalgia pain, by implementing new gamification approaches through the development of an application called FisioQuest. The goal is to enhance patient engagement, encouraging and rewarding their continuous use of the application. To achieve these objectives, it is essential to identify and understand the specific requirements for developing such a mobile application. In the following section, the process and criteria used to determine these requirements are described in detail.

The article is structured as follows: In Section II, the requirements analysis for the FisioQuest application is detailed, outlining both functional and non-functional requirements. Section III defines specific use cases to guide the application's design. Section IV discusses the application design, including the choice of the Flutter framework and the creation of mockups. Finally, Section V addresses the integration of gamification elements to enhance user engagement and outlines plans for usability testing and a co-design process with end-users.

## II. Requirements Analysis

The requirements for the development of the mobile application were determined following a literature review and analysis of case studies [17]-[19]. The research focused on identifying the functionalities and features most valued by patients and most effective in clinical practice. The main points of interest included symptom monitoring techniques, interventions based on physical and mental exercises, and the impact of gamification on treatment adherence. In addition to the scientific literature, existing mobile applications aimed at managing chronic pain were analyzed. This analysis made it possible to identify best practices and the most successful features in terms of effectiveness and engagement, without the need for direct interactions with end users.

Considering this analysis and the application's target audience, the following functional and non-functional requirements were defined.

### A. *Functional Requirements*

- Pain report: Allow users to report, at any time, the presence of pain, indicating the location and level of pain on a 0-10 scale.
- Healthcare specific pages: The application should represent the three types of healthcare professionals who interact with fibromyalgia patients, namely: physicians, nurses and physiotherapists.
- Physician's page: Monitor and record the medication taken daily by the patient.
- Nurse's page: Record and monitor the patient's weight and blood pressure.
- Physiotherapist's page: Recommend suitable physical exercises for fibromyalgia patients and verify their accomplishment.
- Morning/evening questionnaires: Display morning and evening questionnaires on the main page; collect and store user responses for ongoing monitoring.
- Streak system: Implement a streak system to record consecutive days of app utilization and notifies users of their current streak status and daily goals.
- Streak rewards: Allow users to redeem the reward after completing a streak.
- User profile: Display average daily pain statistics in graphical format.

### B. *Non-Functional Requirements*

- Usability: The interface should be intuitive, using basic icons and a minimalist design to reduce cognitive load for FMS patients. Navigation should be simple, allowing quick access to main sections.

- Engagement: Integrate push notifications to remind users of daily tasks and rewards. Use gamification elements to keep users engaged and motivated.
- Performance: The application should load promptly and respond effectively to user interactions.
- Security and privacy: User data must be protected, ensuring compliance with General Data Protection Regulation (GDPR).
- Reliability: The application must minimize crashes and service interruptions. In case of network failure, data must be temporarily stored on the device and synchronized once the connection is restored.
- Compatibility: The application should be compatible with major mobile operating systems (iOS and Android), screen sizes and resolutions.

## III. Use Cases

In order to prototype the user interface, it is important to clearly identify the use cases for the FisioQuest application. To this end, six use cases were defined:

- Pain report - After experiencing an outbreak of pain, the patient opens the pain reporting page. The user selects the location of the pain (from a list) and rates the pain on a scale of 0 to 10. The data is stored in the database and can be viewed in the user's profile in the form of a daily pain graph. Over time, the user and their healthcare provider can track pain patterns and adjust treatments accordingly.
- Daily medication check - A user opens the Physician page to check their medication plan for the day. The page displays a list of medications with checkboxes next to each one. As the user takes their medication, he/she ticks the checkboxes, and the app records the intake. This information is then synchronized with the main screen, showing the "Physician" task as completed for the day.
- Performing an exercise - The user navigates to the physiotherapist's page, where he/she is presented with the list of daily exercises that have been prescribed. Each exercise includes a short video/image and notes on how to properly perform the exercise. The user completes the exercise and marks it as finished. The application records the completion and updates the streak counter.
- Daily questionnaires - In the morning and evening, the user is notified to fill in the daily questionnaires. The Morning Questionnaire asks about the user's sleep quality and morning pain levels, while the Evening Questionnaire asks about general daily pain and how he/she felt throughout the day. The answers are recorded and stored in the database, providing data for monitoring symptoms and identifying potential triggers or patterns.
- Health monitoring - On the Nursing page, users can record their daily weight and blood pressure. They should also mark the daily tasks they have been prescribed, such as "Bathing", "Making the bed", among other activities. This information is stored in the database. Healthcare professionals will be able to observe how weight and blood pressure vary and correlate with pain levels and medication adherence.
- Rewards system - After keeping a series of daily tasks completed for a predefined number of consecutive days, the user receives a notification about their reward. The user can navigate to the rewards section and choose to schedule a teleconsultation with a healthcare professional.

## IV. Application Design

Following the definition of the functional and non-functional requirements, and identifying the use cases, the process of designing the FisioQuest application began by choosing the framework to be used for its development. Considering the non-functional requirements, it was decided to adopt the Flutter framework, which guarantees, among other advantages: the ability to develop a single code base for multiple platforms, such as Android and iOS; the creation of intuitive, visually rich and pleasant user interfaces; high application performance, allowing for fast and responsive navigation/interaction.

### A. *Application Mockups*

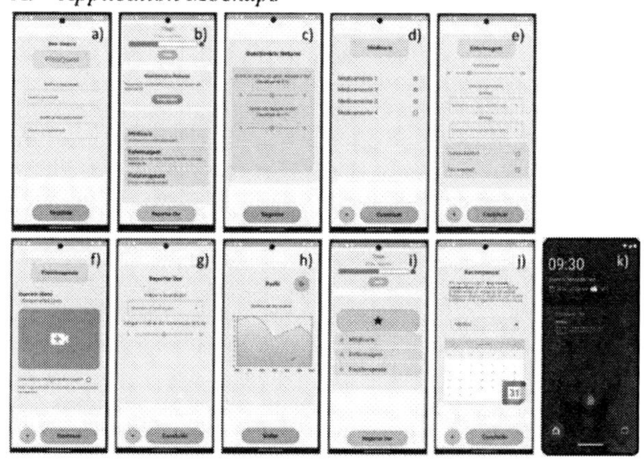

Figure 1. Mockups of the FisioQuest mobile application.

To be able to preview the layout and test the functionality of the application, wireframes and detailed mockups were created which served as models during the development process. The mockups of the main screen of the FisioQuest mobile application are illustrated in Figure 1.

Below is a brief description of the key screens of FisioQuest application:

*a) Login page:* User authentication page.

*b) Main page:* Provides quick access to main features.

*c) Morning / Evening Questionnaire:* Prompts users with questions about sleep quality / morning pain (morning) and daily pain levels and overall condition (evening).

*d) Physician Page:* Allows drug schedules tracking.

*e) Nursing Page:* Allows users to record their daily weight and blood pressure levels.

*f) Physiotherapist Page:* Provides videos, images and notes to assist users in performing their daily exercises.

*g) Pain Reporting Page:* Allows users to report their pain levels and locations.

*h) User Profile Page:* Displays user pain statistics.

*i) Main Page:* Allows users to see their streak status.

*j) Reward Page:* Allows users to claim a reward by scheduling a teleconsultation.

*k) Notification:* Notifies users of pending activities.

### *B. Integrated Gamification Elements*

To encourage user participation and engagement, gamification elements have been integrated into the app. The streak system promotes daily use by tracking usage on consecutive days and notifying users about current streak status and daily goals. When a streak is completed, users can redeem the reward. To make the rewards truly attractive, patients can schedule teleconsultations with health professionals from the Chronic Pain Unit. Instant feedback is given on the user's progress (adherence to medication, reporting of symptoms, physical exercise), thus seeking to reinforce positive behavior. The daily questionnaires (morning and evening) act as regular checks, promoting self-reflection and continuous monitoring of symptoms.

## V. Conclusion and Future Work

Previous studies have pointed to a reduction in pain intensity and an improvement in associated symptoms, such as anxiety and depression, whenever the use of mobile applications was promoted in the approach to chronic pain treatment. The reduction in symptoms with the use of mobile apps can be attributed to the combination of various functionalities, such as performing the recommended physical exercises, better psychotherapeutic support, and improving the quality of data on the patient's condition by continuously monitoring symptoms. These elements are especially important for FMS patients, who often face difficulties in maintaining a self-care routine due to pain and fatigue. It is therefore vital that patients make continued use of the app, and gamification can be a crucial factor.

This paper presents FisioQuest, a mobile app (under development) that incorporates gamification elements to increase patient engagement and facilitate the self-management of pain. The app aims to offer an accessible and convenient solution to complement traditional treatments, reducing the need for pharmacological interventions and their long-term side effects. Additionally, continuous symptom monitoring and recording can enable faster, more effective, and personalized interventions.

The application should initially be subjected to usability tests using the System Usability Scale (SUS) to assess the ease of use of the app and the overall user experience. After the usability testing, a co-design process will be initiated to further enhance the application. This phase will involve the participation of end-users, who will be invited to engage in tests and interviews. Their feedback and insights will play a crucial role in refining the application, ensuring that it meets the needs and expectations of its users effectively. This would be followed by validation through a controlled study, to verify whether the benefits observed were the result of placebo effects or the initial motivation of the participants.

## Acknowledgment

This paper was funded by national funds, through the FCT/MCTES of the projects UIDB/05549/2020 and UIDP/05549/2020.

## References

[1] M. Nicholas et al., "The IASP classification of chronic pain for ICD-11: chronic primary pain. Pain, " Pain, vol. 160, no. 1, pp. 28-37, Jan. 2019, doi: 10.1097/j.pain.0000000000001390.

[2] L. Goudman et al., "Virtual Reality Applications in Chronic Pain Management: Systematic Review and Meta-analysis," JMIR Serious Games, vol. 10, no. 2, pp. e34402, May 2022, doi: 10.2196/34402.

[3] A. S. Gonçalves, F. Monteleone, M. B. Couto, M. E. Machado, and M. E. Pereira, "From automatic thoughts to painful processes: a review on the use of cognitive-behavioral therapies in the management of chronic pain," Revista DOR, vol. 29, no. 1, Nov. 2022, doi: 10.24875/DOR.22000003.

[4] D. Teater, The psychological and physical side effects of pain medication, National Safety Council, 2015. [Online]. Available from: https://www.nsc.org/getmedia/0113f259-d2c5-4a3e-abca-f05299f65ec2/adv-rx-side-effects-wp.pdf 2024.09.26

[5] M. D. Cheatle, "Prescription opioid misuse, abuse, morbidity, and mortality: balancing effective pain management and safety," Pain Med., vol. 16, suppl. 1, pp. S3-S8, Oct. 2015, doi: 10.1111/pme.12904. PMID: 26360931.

[6] S. L. K. Yuan, L. A. Couto, and A. P. Marques, "Effects of a six-week mobile app versus paper book intervention on quality of life, symptoms, and self-care in patients with fibromyalgia: a randomized parallel trial," Braz J Phys Ther., vol. 25, no. 4, pp. 428-436, Jul-Aug 2021, doi: 10.1016/j.bjpt.2020.10.003.

[7] F. Wolfe et al., "The American College of Rheumatology preliminary diagnostic criteria for fibromyalgia and measurement of symptom severity," Arthritis Care Res (Hoboken), vol. 62, no. 5, pp. 600-10, May 2010, doi: 10.1002/acr.20140. PMID: 20461783.

[8] P. Sarzi-Puttini et al., "Fibromyalgia: an update on clinical characteristics, aetiopathogenesis and treatment," Nat Rev Rheumatol, vol. 16, pp. 645–660, 2020, doi: 10.1038/s41584-020-00506-w.

[9] R. de la Vega, R. Roset, S. Galán, and J. Miró, "Fibroline: A mobile app for improving the quality of life of young people with fibromyalgia," Journal of Health Psychology, vol. 23, no. 1, pp. 67-78, 2018, doi: 10.1177/1359105316650509.

[10] S. L. K. Yuan and A. P. Marques, "Development of ProFibro - a mobile application to promote self-care in patients with fibromyalgia," Physiotherapy, vol. 104, no. 3, pp. 311–317, 2018, doi: 10.1016/j.physio.2018.04.005.

[11] J. An, W. Fan, A. Mittal, Y. Zhang, and A. T. Chen, "Mobile App Use among Persons with Fibromyalgia: A Cross-sectional Survey," J Pain, Mar. 24, 2024, doi: 10.1016/j.jpain.2024.03.011.

[12] J. Miró, M. Lleixà-Daga, R. de la Vega, P. Llorens-Vernet, and M. P. Jensen, "A Mobile Application to Help Self-Manage Pain Severity, Anxiety, and Depressive Symptoms in Patients with Fibromyalgia Syndrome: A Pilot Study," Int. J.

Environ. Res. Public Health, vol. 19, no. 19, p. 12026, 2022, doi: 10.3390/ijerph191912026.

[13] N. X. Thanh et al., "Economic burden of chronic pain in Alberta, Canada," PLoS ONE, vol. 17, no. 8, p. e0272638, 2022, doi: 10.1371/journal.pone.0272638.

[14] L. Barham, "Economic Burden of Chronic Pain Across Europe," Journal of Pain & Palliative Care Pharmacotherapy, vol. 26, no. 1, pp. 70–72, 2012, doi: 10.3109/15360288.2011.650364.

[15] T. Takura et al., "The societal burden of chronic pain in Japan: an internet survey," J Orthop Sci, vol. 20, pp. 750–760, 2015, doi: 10.1007/s00776-015-0730-8.

[16] L. F. Azevedo et al., "The economic impact of chronic pain: a nationwide population-based cost-of-illness study in Portugal," Eur J Health Econ, vol. 17, pp. 87–98, 2016, doi: 10.1007/s10198-014-0659-4.

[17] M. Cascella et al., "Envisioning gamification in anesthesia, pain management, and critical care: basic principles, integration of artificial intelligence, and simulation strategies," J Anesth Analg Crit Care, vol. 3, p. 33, 2023, doi: 10.1186/s44158-023-00118-2.

[18] A. S. Miller, J. A. Cafazzo, and E. Seto, "A game plan: Gamification design principles in mHealth applications for chronic disease management," Health Informatics Journal, vol. 22, no. 2, pp. 184-193, 2016, doi: 10.1177/1460458214537511.

[19] L. Sardi, A. Idri, and J. L. Fernández-Alemán, "A systematic review of gamification in e-Health," Journal of Biomedical Informatics, vol. 71, pp. 31-48, Jul. 2017, doi: 10.1016/j.jbi.2017.05.011.

# Transforming Healthcare: The Role of AI and Informatics in Modern Medical Support

Amit Srivastava
International Organization for Academic and Scientific Development (IOASD)
Baltimore, USA
email: amit14321@gmail.com

*Abstract*—**This paper explores the transformative role of Artificial Intelligence (AI) and informatics in healthcare, focusing specifically on advancements in diagnostics, personalized treatment strategies, data security, and regulatory frameworks. AI has significantly improved patient management and treatment outcomes through applications such as predictive analytics, AI-driven genomics, and enhanced diagnostic imaging. The study also discusses key challenges in data privacy, the ethical implications of AI, and the integration of these technologies within healthcare infrastructure. Future directions highlight the potential of AI in genomics, telemedicine, and drug discovery, underscoring the need for robust frameworks to guide ethical and safe implementation.**

***Keywords—AI; informatics; healthcare; diagnosis; treatment.***

## I. INTRODUCTION

Artificial Intelligence (AI) is revolutionizing healthcare by enhancing diagnostic accuracy, optimizing treatments, and supporting overall patient care. Leveraging large datasets, AI applications offer new insights into patient conditions, aiding healthcare professionals in predictive analysis and clinical decision-making. For instance, AI algorithms have been used to predict cancer treatment outcomes, assess cardiovascular risks, detect diabetic retinopathy in imaging, and even forecast disease progression in chronic conditions [1]. The broad application of AI in healthcare illustrates its potential to analyze both structured and unstructured data, providing crucial support in improving patient outcomes across diverse domains. This study examines the transformative role of AI in healthcare, focusing primarily on its impact on diagnostics, personalized treatment, and healthcare informatics. Specifically, it explores how AI-driven data analysis can identify and prevent healthcare issues proactively and how AI applications enhance patient care through improved diagnostics, treatment personalization, and operational efficiency in healthcare systems. The study also discusses AI's integration into electronic health records, health information exchange, and clinical decision support systems. Future directions highlight areas such as AI's potential in genomics for precision medicine. Data from medical records, wearable devices, and diagnostic imaging tools offer comprehensive insights for risk assessment, predictive analysis, and early intervention [1]. However, the integration of AI in healthcare raises significant challenges. Key concerns include privacy issues in managing sensitive health data, necessitating robust security measures to ensure compliance with regulatory standards [2]. Additionally, there is a critical need for interpretable AI models that healthcare providers and patients can easily understand, fostering trust and accessibility.

The remainder of this paper is organized as follows. Section II discusses the role of AI in diagnostics and personalized treatment. Section III explores the integration of informatics in modern medical support, particularly focusing on electronic health records, health information exchange, and clinical decision support systems. Section IV examines the benefits of AI, while Section V delves into challenges and ethical considerations. In Section VI, real-world applications of AI are highlighted through a case study; Section VII reviews policy and regulatory implications. Section VIII presents future directions, including advancements in genomics, telemedicine, and drug discovery. Finally, Section IX provides concluding remarks.

## II. THE ROLE OF AI IN HEALTHCARE

### A. Diagnostics

Advancements in artificial intelligence have significantly improved the accuracy and efficiency of diagnostics, particularly in medical imaging. AI algorithms are now used extensively in various imaging modalities, including X-ray, Computed Tomography (CT) scans, and Magnetic Resonance Imaging (MRI), where they help detect conditions such as fractures, tumors, and neurological disorders. Esteva et al. [3] highlighted AI's effectiveness by demonstrating how AI systems can diagnose skin cancer with accuracy comparable to dermatologists, identifying abnormalities that may be overlooked by human observation [3]. AI tools such as Aidoc and Zebra Medical Vision assist radiologists by pre-screening images for potential issues, thus speeding up the diagnostic process and allowing for quicker interventions.

In addition to these applications, specific software solutions have been developed to enhance image quality and aid in diagnosis. For example, DeepMedic is used for tumor segmentation in MRI, and VoxelMorph improves MRI image registration, which is crucial for comparing images taken over time to monitor disease progression. Similarly, in CT imaging, V-Net is used to enhance segmentation accuracy, particularly in detecting lung and liver cancers. The combination of advanced algorithms with imaging modalities like CT scans and MRIs ensures that complex issues, such as cancer and neurological disorders, can be identified more accurately and efficiently than with traditional methods [1]. These AI-powered tools illustrate the transformative role of AI in diagnostics, contributing to improved patient outcomes and optimized clinical workflows.

### *B. Personalized Treatment*

AI technology has enabled the development of highly individualized treatment plans by analyzing vast datasets, including genetic information and lifestyle factors. For instance, in oncology, AI-driven platforms like IBM Watson for Oncology and Foundation Medicine analyze clinical literature, patient medical records, and genetic profiles to recommend targeted therapies based on each patient's unique genetic mutations. AI methods have shown promise in predicting chemotherapy response and tailoring treatment strategies to minimize side effects, enhancing the effectiveness of treatments. By employing precision medicine techniques, healthcare providers can reduce adverse reactions while improving overall outcomes [2].

In cardiology, AI helps predict patient responses to medications based on genetic and lifestyle data, allowing for adjustments that optimize treatment efficacy and reduce the risk of complications. Studies have shown that using AI for pharmacogenomic analysis—identifying genetic markers associated with medication responses—enables the personalization of treatments for chronic conditions like hypertension and diabetes [4]. These AI-driven advancements in personalized treatment highlight the power of data analysis in tailoring therapies, leading to safer, more effective healthcare.

## III. Informatics in Modern Medical Support

### *A. Electronic Health Records (EHRs)*

Electronic Health Records (EHRs) have transformed healthcare operations by providing immediate access to comprehensive patient data, improving decision-making and patient outcomes. AI integration into EHR systems enhances these benefits by enabling predictive analysis and early detection of health risks. For example, AI models analyze patient history and lab results to predict potential complications for patients with chronic illnesses, such as diabetes or heart disease, allowing providers to intervene early. A study demonstrated how AI-enhanced EHRs can reduce human error in record-keeping by flagging abnormal results or highlighting patient histories that may need closer monitoring [1]. Some AI-powered EHR platforms, such as Epic Systems and Cerner Health, now incorporate machine learning algorithms to support personalized care recommendations, streamline workflows, and reduce manual data entry errors.

### *B. Health Information Exchange (HIE)*

Health Information Exchange (HIE) systems facilitate the secure sharing of patient data across various healthcare providers, ensuring continuity of care regardless of geographic or institutional boundaries. AI's role in HIE's includes data standardization, which is essential for making patient information usable and comparable across systems. AI-driven platforms like MedRec use blockchain and machine learning to secure and streamline data exchange between providers, improving the availability of real-time patient information. Obermeyer and Emanuel [5] emphasized that such interoperability allows physicians to access comprehensive patient histories, reducing duplicate tests and unnecessary treatments. Moreover, AI-enhanced HIEs improve data retrieval speeds, enabling providers to make faster, data-driven decisions, particularly in emergency or critical care settings.

### *C. Clinical Decision Support Systems (CDSS)*

Clinical Decision Support Systems (CDSS) utilize AI to provide healthcare providers with intelligent, data-driven insights to aid in diagnosis and treatment planning. AI-powered CDSS tools, such as IBM Watson Health and MDClone, offer diagnostic suggestions based on the latest clinical research, patient-specific data, and real-time analysis of current symptoms. For instance, these systems can analyze a patient's symptoms and suggest potential diagnoses or flag treatments that may cause adverse reactions based on the patient's medical history. Such tools are precious in complex cases where multiple comorbidities are involved, as they help providers consider a broader range of diagnostic possibilities and treatments. According to studies, AI-driven CDSS systems have improved diagnostic accuracy rates by up to 30%, helping to reduce misdiagnosis and ensure that patients receive appropriate and timely care [1].

## IV. Benefits of AI and Informatics in Healthcare

### *A. Improved Accuracy and Efficiency*

The precision with which AI can interpret massive datasets minimizes human error significantly. AI-driven technologies evaluated radiological data more quickly than human radiologists in research by Jiang et al. (2017), and they achieved accuracy rates of more than 90% [4]. This accuracy and speed contribute to improved diagnostic results and more effective workflow in the healthcare industry. The accuracy rates of several AI-driven diagnosis systems for imaging tests, including MRIs, CT scans, and X-rays, are contrasted in the following graph. AI constantly performs better than human diagnosis, which improves the early identification of serious illnesses like cancer.

### *B. Enhanced Patient Outcomes*

Artificial Intelligence (AI) and informatics technologies significantly improve patient outcomes by enabling early disease detection, personalized treatment plans, and real-time monitoring. AI-driven tools facilitate earlier identification of diseases, as demonstrated in cancer diagnostics. For example, the Google DeepMind algorithm has been used in clinical settings to detect early signs of breast cancer in mammograms with greater accuracy than human radiologists, allowing for earlier intervention and improved survival rates [4].

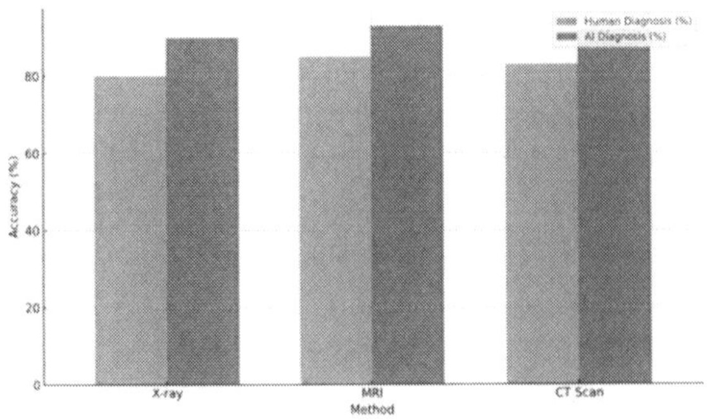


Figure 1. Comparison of Diagnostic Accuracy: Human vs AI.

In managing chronic diseases, AI has shown positive results, particularly in diabetes care. AI-enhanced Continuous Glucose Monitoring Systems (CGMS), such as those integrated with platforms like Dexcom and Medtronic Guardian Connect, help patients with diabetes by providing continuous, real-time glucose monitoring. These systems predict blood glucose trends based on historical data and activity levels, allowing patients and caregivers to make timely adjustments in insulin dosages, thereby reducing hypoglycemic events by as much as 40% and improving long-term HbA1c levels [6].

Furthermore, AI applications in heart disease management have improved outcomes by predicting patient risk of cardiac events. Platforms like HeartFlow use AI-based image analysis to assess coronary artery blockages non-invasively, which aids cardiologists in treatment planning. Studies indicate that using such AI tools has reduced unnecessary invasive procedures by up to 30%, allowing patients to receive optimal care while minimizing risk [4]. By facilitating personalized, real-time healthcare interventions, AI continues to play a pivotal role in enhancing patient outcomes across various conditions.

TABLE I. COST SAVINGS IN HEALTHCARE DUE TO AI.

| Country/Region | Healthcare Sector | AI Application Area | Estimated Cost Savings (in billions USD) | Primary Sources of Savings |
|---|---|---|---|---|
| United States | Diagnostics | Imaging & Radiology | 15 | Reduced misdiagnosis, faster diagnostic workflows |
| Europe | Chronic Disease Management | Predictive Analytics | 10 | Fewer hospitalizations, preventive management |
| Asia-Pacific | Personalized Treatment | Oncology | 8 | Reduction in adverse reactions |
| North America | Operational Efficiency | Hospital Resource Planning | 7 | Optimized staffing, reduced downtime |
| Global | Drug Discovery | Pharmaceutical Development | 20 | Shortened R&D timelines, decreased overhead |

### C. Cost Reduction

AI and informatics can help reduce healthcare costs by streamlining operations, reducing the need for unnecessary tests and procedures, and improving resource allocation. This efficiency can lead to significant cost savings for healthcare providers and patients. The table below demonstrates the cost savings achieved in various areas of healthcare through AI implementations. These figures are based on industry reports and studies highlighting the economic benefits of using AI in diagnostics, predictive analytics, and personalized treatment.

Table I summarizes the estimated cost savings across different areas of healthcare where AI has been implemented. The industry research and studies data highlight how AI helps streamline processes and reduce costs [7].

## V. CHALLENGES AND ETHICAL CONSIDERATIONS

While artificial intelligence has the capacity to revolutionize healthcare, it also introduces considerable obstacles and ethical quandaries. A key issue is data security and privacy. In 2015, a significant cyberattack targeted Anthem Inc., a leading U.S.-based health insurance provider headquartered in Indianapolis, Indiana. The breach compromised the personal health information of approximately 79 million individuals, exposing names, birth dates, Social Security numbers, addresses, and employment information. This incident underscored the vulnerabilities in healthcare data security and highlighted the urgent need for enhanced cybersecurity measures to protect sensitive patient information. AI algorithms, frequently trained on extensive datasets, are susceptible to biases. A 2019 study revealed that an AI system employed by U.S. hospitals was less inclined to send black patients for specialist care than white patients despite comparable health problems. Unaddressed prejudices can sustain disparities in healthcare provision [8].

Furthermore, the transparency of artificial intelligence systems becomes an additional ethical consideration. Numerous AI models function as "black boxes," complicating the comprehension of decision-making processes for healthcare providers and patients. Ensuring the interpretability and explainability of AI tools is essential for preserving trust in the healthcare system. The European Union's General Data Protection Regulation (GDPR) underscores the "right to explanation," requiring that patients be informed about the processes underlying AI-driven decisions in their healthcare. Healthcare practitioners must implement stringent data protection procedures and promote transparent, equitable AI systems that respect patient rights and ethical standards.

Real-world applications of AI and informatics in healthcare demonstrate their transformative potential. Case studies showcasing successful implementations can provide valuable insights and best practices for other healthcare organizations looking to adopt these technologies.

*Case Study: AI in Closed-Loop Continuous Glucose Monitoring Systems (CGMS):*

Developing Closed-Loop Continuous Glucose Monitoring Systems (CGMS) has been one of the most revolutionary uses of AI in healthcare. By combining AI-driven insulin pumps with continuous glucose monitoring, these devices create an "artificial pancreas" that automatically regulates insulin levels. Artificial intelligence algorithms forecast blood glucose levels several hours in advance by utilizing CGMS data as well as outside variables like physical activity and meal consumption. By examining past data and trends, the AI system can maximize insulin delivery and reduce the risk of hyperglycemia and hypoglycemia. AI-enhanced Continuous Glucose Monitoring Systems (CGMS) have significantly improved glucose management for patients with type 1 diabetes globally. According to a study conducted in the United States and published in *The New England Journal of Medicine by the Juvenile Diabetes Research Foundation*, these devices reduced hypoglycemia incidents by up to 33% to 50%, as they predict glucose trends and adjust insulin delivery in real-time [9]. These devices assess and forecast glucose trends based on real-time sensor data using machine learning algorithms, which enables better blood glucose control and prompt treatments. These developments improve the quality of life for diabetic patients by giving them a new degree of independence and convenience.

## VI. Policy and Regulatory Implications

Policy and regulatory frameworks that support the use of AI and informatics in healthcare are essential for ensuring the ethical and safe application of these technologies. Governments and regulatory agencies play a critical role in setting rules that guarantee patient safety, privacy, and equitable access to AI-driven healthcare solutions. These frameworks need to address key concerns such as data security, transparency in AI decision-making, and the mitigation of biases in algorithms. Regulatory bodies like the U.S. Food and Drug Administration (FDA) have already begun setting guidelines for AI technologies in healthcare, such as developing the Software as a Medical Device (SaMD) framework, which provides a pathway for the approval and monitoring of AI tools in clinical settings [10]. Furthermore, the European Union has introduced the General Data Protection Regulation (GDPR), which enforces strict standards for data privacy and security, particularly when dealing with sensitive healthcare data. AI systems in healthcare must comply with GDPR rules regarding patient consent and the right to access information about how AI decisions are made. These regulations are essential to maintaining patient trust and preventing misuse of personal health data. Current efforts to enhance AI regulation also include initiatives like the Artificial Intelligence Initiative Act in the U.S., which seeks to establish a national strategy for AI and provide funding for research into AI ethics and safety [11]. Similarly, the European Union's Artificial Intelligence Act aims to establish a risk-based framework to regulate AI, with healthcare applications classified as high-risk, requiring stringent oversight and compliance with ethical standards [12].

Looking ahead, policymakers must encourage innovation by creating regulatory sandboxes that allow AI developers to test and refine their algorithms in controlled environments while maintaining safety standards. Governments must also foster collaboration between public and private sectors to promote ethical AI development while ensuring that patient's rights are upheld. By establishing robust frameworks, regulatory bodies can facilitate the responsible growth of AI in healthcare, improving patient outcomes and driving the next generation of medical advancements.

## VII. Future Directions

### *A. Integration of AI and Genomics*

The capacity to transform medicine is significant with the amalgamation of AI and genomics. The integration of AI in genomic data analysis holds the potential to reveal insights into the genetic origins of diseases, facilitating the development of personalized and precise treatment strategies for individuals. AI-driven frameworks excel at identifying signs associated with diseases, enabling healthcare providers to foresee susceptibility to illnesses, treatment responses, and potential harmful effects. Artificial intelligence can be employed to discern patterns in an individual's genetic composition that may indicate a heightened risk of acquiring various cancers and other hereditary disorders, facilitating early preventive interventions. For instance, Visibelli et al. [13] highlight AI's potential to accelerate genomic data interpretation and AI's role in analyzing complex genetic data to tailor healthcare based on individual genetic profiles. The study emphasizes unsupervised learning models in discovering disease biomarkers and advancing patient-specific treatment strategies. Additionally, AI in genomics possesses the capacity to optimize the interpretation process by conserving time and reducing complications. The conventional methodology for genomic analysis is labor-intensive. Demands specialist expertise. Nonetheless, AI programs can rapidly evaluate this data. Furnish healthcare practitioners with information that can augment treatment efficacy. With the increasing prevalence of genetic testing, the amalgamation of AI and genomics will persist in advancing precision medicine, yielding enhanced healthcare solutions customized for people worldwide.

### *B. Expansion of Telemedicine*

The expansion of AI-driven telemedicine systems is diminishing barriers to healthcare access for patients in urban and rural areas. Facilitating a beneficial transformation in the healthcare sector. These platforms utilize artificial intelligence to provide diagnostic and therapy recommendations while monitoring health issues. They are offering essential assistance to folks who may struggle to obtain timely medical care otherwise. Research by Keesara et al. [14] in *The New England Journal of Medicine* discusses how telemedicine enhances access to healthcare services, noting that AI-driven platforms have reduced patient wait times and eliminated geographical barriers.

Telemedicine may leverage AI's real-time patient data analysis to deliver customized care plans and continuously assess patient health. Implement requisite modifications to treatments as necessary. Healthcare practitioners gain advantages from AI-driven telemedicine, as it reduces the necessity for physical visits by enabling checkups, consultations, and follow-up appointments. This enhances healthcare accessibility. Additionally, it mitigates healthcare disparities by delivering uniform quality care to all patients, irrespective of their geographical location. The expansion of telemedicine contributes to addressing the healthcare disparities faced by communities with limited access to specialized medical services. As telemedicine advances, AI is poised to facilitate the development of tools and virtual therapies, delivering equitable healthcare solutions to numerous persons.

### *C. AI in Drug Discovery*

AI is quickly changing how new drugs are found by speeding up the process of identifying drug options and forecasting their effectiveness accurately. Traditional methods for discovering drugs require a lot of work and money. It can take years to produce outcomes. On the other hand, AI can analyze data from molecular research studies and patients' and real-life experiences to reveal trends and forecast which substances could work well against particular illnesses. For example, Google introduced Alpha Missense, which assesses genetic variant impacts on diseases. This tool could significantly improve early-stage drug discovery by guiding researchers to potential targets. A study by Tordai et al. [15] highlights AlphaFold's applications in understanding protein structures and screening potential drugs, helping scientists quickly identify molecules with high therapeutic promise. Through the utilization of machine learning algorithms, AI has the ability to replicate systems, simulate interactions between drugs, and pinpoint potential candidates at a significantly accelerated pace compared to traditional techniques. Moreover, AI has the capability to enhance all aspects of the drug discovery process, from pinpointing targets to conducting trials. Its capacity extends to scrutinizing the composition of drugs, predicting adverse reactions, and pinpointing already established medications that may be repurposed for novel therapeutic applications. This streamlining potential in drug discovery could potentially bring down costs of development, enforce timelines, and culminate in expedited approvals for treatments. As artificial intelligence progresses further into the future, we can expect advancements in fields like diseases, where conventional methods of drug development have frequently struggled due to the intricate nature and limited number of patients involved.

## VIII. Implementation Strategies

### A. Infrastructure requirements

Implementing AI and informatics in healthcare requires a robust and scalable technological infrastructure. This includes not only high-performance computing resources but also secure and flexible data storage solutions, which are critical for managing the vast amounts of patient data used in AI applications. Healthcare organizations must also ensure reliable and high-speed network connectivity to support real-time data processing and cloud-based AI solutions. Additionally, advanced cybersecurity measures must be put in place to protect sensitive patient information from potential breaches. Investment in scalable and interoperable systems will be necessary to accommodate future AI advancements and the growing demand for healthcare data analytics. Infrastructure should also be adaptable to integrate new technologies without disrupting ongoing operations.

### B. Workforce training and education

Integrating AI and informatics into healthcare also depends on equipping the workforce with the knowledge and skills to use these technologies effectively. Healthcare professionals need comprehensive training not only on the technical aspects of AI tools but also on the ethical and legal implications, such as patient data privacy and the responsible use of AI in clinical decision-making. Continuous education programs, including workshops, certification courses, and simulations, can help bridge the existing knowledge gap. Furthermore, interdisciplinary collaboration between IT experts, healthcare professionals, and data scientists will be crucial in ensuring the effective deployment of AI systems. Encouraging a culture of innovation and adaptability within healthcare organizations will also be essential for integrating these new technologies smoothly. Additionally, fostering collaboration between healthcare professionals and AI experts will ensure a smoother integration of AI technologies, allowing for better adoption and innovation.

## IX. Conclusion and Future Work

As we have seen, the integration of informatics and Artificial Intelligence (AI) in healthcare leads to notable improvements in patient management, diagnosis, tailored therapy, and operational effectiveness. With predictive analytics and customized treatment plans, AI's accurate processing of massive datasets enhances diagnostic precision and allows for more individualized care. These innovations improve patient outcomes, save expenses, and streamline healthcare procedures. This paper presents research that demonstrates the revolutionary potential of Artificial Intelligence (AI) across a range of medical applications, such as predictive analytics, Electronic Health Records (EHRs), and diagnostics. AI and informatics have the potential to completely transform the way that healthcare is delivered by streamlining workflows and decision-making procedures and making them more accurate, patient-centered, and efficient. Future research in this area should focus on ethical issues, security concerns, and data privacy, especially as they relate to AI algorithms. Robust frameworks will be necessary for future developments in AI-driven healthcare solutions to guarantee accuracy, fairness, and transparency. Furthermore, as AI has the potential to significantly advance the fields of genetics, telemedicine, and drug discovery, the continued development of AI tools must remain focused on these real-world uses. Looking ahead, to guarantee the practical and moral application of AI technology, it is critical to build solid partnerships between technologists, legislators, and healthcare professionals. AI has a bright future in healthcare, with the potential to increase its impact on bettering patient outcomes, cutting costs, and raising standards of care globally.

## References


[1] A. Rajkomar, J. Dean, and I. Kohane, "Machine learning in medicine," The New England Journal of Medicine, vol. 380, pp. 1347-1358, Apr. 2019, doi:10.1056/NEJMra1814259.

[2] E. J. Topol, "High-performance medicine: The convergence of human and artificial intelligence," Nature, vol. 25, pp. 44-56, Jul. 2019, doi:10.1038/s41591-018-0300-7.

[3] A. Esteva et al., "Dermatologist-level classification of skin cancer with deep neural networks," Nature, vol. 542, pp. 115-118, Jan. 2017, doi:10.1038/nature21056.

[4] F. Jiang et al., "Artificial intelligence in healthcare: Past, present and future," Stroke and Vascular Neurology, vol. 2, pp. 230-243, Dec. 2017, doi:10.1136/svn-2017-000101.

[5] Z. Obermeyer and E. J. Emanuel, "Predicting the future — Big data, machine learning, and clinical medicine," The New England Journal of Medicine, vol. 375, pp. 1216-1219 Sep. 2016, doi:10.1056/NEJMp16061.

[6] The Juvenile Diabetes Research Foundation Continuous Glucose Monitoring Study Group, "Continuous Glucose Monitoring and Intensive Treatment of Type1 Diabetes," The New England Journal of Medicine, vol. 380, pp. 1464-1476, Oct. 2008, doi:10.1056/NEJMoa0805017.

[7] Frost & Sullivan: Global artificial intelligence in healthcare market report.[Online]. Available from: https://www.frost.com/news/press-releases/artificial-intelligence-healthcare-takes-precision-medicine-next-level.

[8] Z. Obermeyer, B. Powers, C. Vogeli, and S. Mullainathan, "Dissecting racial bias in an algorithm used to manage the health of populations," Science, vol.366, pp. 447-453, Oct. 2019, doi:10.1126/science.aax2342.

[9] T. Davenport and R. Kalakota, "The potential for artificial intelligence in healthcare," Future Healthcare Journal, vol. 6, pp. 94-98, Jun. 2019, doi:10.7861/futurehosp.6-2-94.

[10] U.S. Food and Drug Administration: Artificial Intelligence/Machine Learning (AI/ML)-Based Software as a Medical Device (SaMD) Action Plan [Online] Available from: https://www.fda.gov.

[11] Artificial Intelligence Initiative Act (2020):[Online] Available from: https://www.congress.gov/bill/116th-congress/senate-bill/1558.

[12] European Commission: Artificial Intelligence Act: [Online] Available from:https://eur-lex.europa.eu/legal-content/EN/TXT/?uri=CELEX%3A52021PC0206.

[13] A. Visibelli, B. Roncaglia, O. Spiga, and A. Santucci, "Unsupervised Learning in Precision Medicine: Unlocking Personalized Healthcare through AI," Applied Sciences, vol. 14, no. 20, pp. 9305, Oct. 2024, doi:10.3390/app14209305.

[14] S. Keesara, A. Jonas, and K. Schulman, "COVID-19 and Health Care's Digital Revolution," The New England Journal of Medicine, vol. 382, pp. e82, Apr. 2020, doi:10.1056/NEJMp2005835.

[15] H. Tordai, O. Torres, M. Csepi, R. Padányi, G. L. Lukács, and T. Hegedűs, "Analysis of AlphaMissense data in different protein groups and structural context," Nature, vol.11, pp. 495, Apr. 2024, doi:10.1038/s41597-024-03327-8.

# Social Engagement Embeddings of Parkinson's Disease through Autoencoders

Ting Xiao
*Department of Information Science*
*University of North Texas*
Denton, USA
email: ting.xiao@unt.edu

Poulomi Guha
*Department of Computer Science*
*University of North Texas*
Denton, USA
email: poulomiguha@my.unt.edu

Ali Shah Tajuddin
*Department of Computer Science*
*University of North Texas*
Denton, USA
email: alishahlnu@my.unt.edu

***Abstract*—Various survey tools are available to measure social engagement, but they often suffer from infrequent measurement and recall bias. To address this, we developed a mobile application that estimates turn-taking in conversations and generates engagement features. These features were used to create an autoencoder-based hidden representation of individuals, which distinguishes between Parkinson's Disease and control subjects. The study aims to create reduced representations to robustly compare speaker-test outcomes with limited samples. An autoencoder was employed to reduce the number of features related to social engagement. This tailored assessment tool was applied to extract 42 speaker assessment scores, which were distilled into two-dimensional embeddings using a 9-layer autoencoder. We compared the proposed hidden representation with Principal Component Analysis, assessing metrics such as conversation percentage, turn-taking, and total pauses. These embeddings enabled a cross-validated reconstruction of all 42 features, accounting for 58% of the variance and were validated using multiple classification methods, including K-Nearest Neighbors (KNN), Support Vector Machine, Random Forest, and XGBoost. The KNN model, using the embeddings features, achieved a 90% macro precision score. Our results suggest that autoencoder representations provide a concise and effective tool for the holistic assessment of speaker behavior in limited data scenarios.**

***Index Terms*—*Autoencoder*; *Parkinson's Disease*; *Hidden Representation*; *Mobile Application*.**

## I. INTRODUCTION

Communication disorders affect 5% to 10% of the US population [1][2]. Individuals affected by these disorders exhibit impaired abilities in listening, speaking, writing, reading, and social interaction. The benchmark for rehabilitation is that these individuals must achieve the ability to communicate effectively and independently in natural environments following hospital discharge [3]. However, the treatment of cognitive communication issues are challenging due to the high variability in therapeutic approaches required for different patients. Notably, the potential disability of everyday communication effectiveness and independence termed Participation Restriction cannot be predicted from the nature or severity of the underlying speech or language impairment as assessed in the hospital [4][5]. Furthermore, to reduce financial expenditures, medical practitioners should focus on most outcomes in the shortest period of time due to the short duration of inpatient and outpatient rehabilitation. Even after in-hospital rehabilitation, residual impairments and associated participation restrictions are common and may last a lifetime [6], which may result in negative social outcomes [7] and emotional problems [8] for the person with the communication disorder. Thus, a valid assessment for potential communicative participation restriction is essential to augment long-term health outcomes and patient quality of life while reducing societal costs.

Language impairment is difficult to assess in neurodegenerative conditions such as Parkinson's Disease (PD) and Alzheimer's Disease (AD). There have been few studies on how to measure individuals with communication difficulties in their natural setting, such as at home, and the evaluation methods used can be highly diverse, such as post-hoc and point-in-time self-report scales [9][10]. As a result, bridging the gap is a significant issue for researchers. Additionally, speech evaluations require detailed protocols and are influenced by factors like geographic location and physician availability. Fortunately, as technology has advanced in at-home devices, cell phone applications to monitor speech have been developed [11][12]. Mobile phones are affordable and accessible and have enormous potential to assess speech [13][14]. In this study, we investigated the use of speech and language features to distinguish between PD and control subjects and developed a mobile application designed to easily capture these features.

Communication impairment is a prevalent symptom of PD. Research demonstrates varied impacts: Lang et al. highlighted impaired pragmatic communication in PD [15], Robinsons et al. observed reduced spontaneity [16], and Dushanova et al. linked changes in verb and noun usage to the disease [17]. Despite these studies, there is no automated globally approved scale that considers speech and language features. This gap has spurred the use of data-driven deep learning techniques to refine speech assessment. For example, Yeung et al.'s analysis of speech characteristics confirmed clinician agreement on features such as word finding difficulties, which correlate with the number of pauses, word duration, and syntactic complexity—key indicators of language impairment [17][18]. Orozco-Arroyave evaluated nonlinear dynamics features and showed that up to 76.81% accuracy could be achieved using the utterance of vowels [19]. Berus et al. achieved an accuracy of 86.47% by applying multiple feature selection processes and proposed feedforward artificial neural networks to classify PD [20]. Additionally, Tsanas et al. developed a machine learning model integrated system that assesses both PD subjects and the quality of their speech [21]. These studies indicate the potential of using deep learning algorithms to assist clinicians in accurately diagnosing PD. In this essence, we adopted a data-driven approach by extracting speech and language features to accurately identify PD.

Clinicians rely on a variety of clinical measures, many of which are collected by automated systems and used for judgment. However, the abundance of individual metrics can create challenges in synthesizing information and drawing clear conclusions. To overcome this issue, lowering the size of the data and developing summary metrics can help. Conventional strategies for reducing dimensions include Principal Component Analysis (PCA) [22] and Linear Discriminant Analysis (LDA). However, these linear approaches fail to capture the non-linear correlations between input data, resulting in a less efficient summary representation. Recently, neural network techniques such as autoencoders have shown promise in learning lower representations of high-dimensional data for natural language processing and computer vision tasks. AutoEncoders (AEs) are neural network algorithms that learn hidden representations of high-dimensional data [23][24].

Previous works have shown that AEs can learn meaningful hidden representations that lead to better downstream tasks, including Ng et al. using speech signals and demonstrated autoencoder-based representation to distinguish disordered speech PD [25]. In order to correctly classify PD by accurate interpretation of the speech and vocal data, Caliskan et al. suggested stacked autoencoder embeddings [26]. Gunduz et al. used a convolution neural network with a vocal feature set to differentiate between PD and control subjects [27]. Hoq et al. compared several models to classify PD and control subjects. Based on a Support Vector Machine (SVM) [28] integrated with a Principal Component Analysis (PCA) and a Sparse AutoEncoder (SAE), the results demonstrated that the proposed SAE-SVM model outperformed not only the PCA-SVM and standard models such as Multilayer Perceptron (MLP), XGBoost [29], K-Nearest Neighbors (KNN), and Random Forest (RF), but also surpassed two recent studies on the same dataset. Additionally, performance was further enhanced by applying SMOTE for oversampling and dataset balancing [30].

Using a data-driven approach will enhance early diagnosis and reduce the amount of time for the diagnostic process. For this project, we developed a prototype mobile application that simplifies the extraction of necessary attributes. We utilized an AutoEncoder (AE) to learn the hidden representations of subjects, classifying between PD and control using embeddings in KNN, SVM, RF, and XGBoost. Most importantly, we demonstrated that hidden representations could effectively capture the full spectrum of an individual's social participation, proving that AE is a robust method for generating meaningful representations of individuals.

The rest of the paper is structrued as follows. In Section II, we present the materials and methods used. The results are shown and discussed in Section III. We conclude the work in Section IV.

## II. MATERIALS AND METHODS

### A. Study Design

We have four modules in this study, as shown in Figure 1. The first two modules are designed to extract features efficiently, while the last two demonstrate how these features can be utilized in downstream analysis.

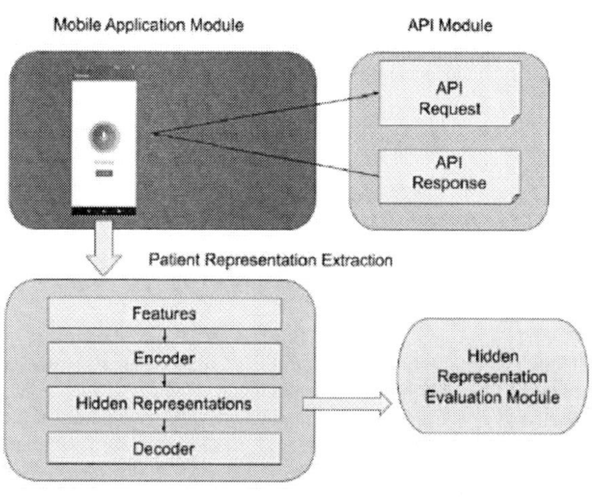


Figure 1. Study Design: Four modules: 1) API, 2) Mobile Application, 3) Patient Representation Extraction, and 4) Evaluation.

1) API: We have built a web API using node.js and Google Speech-to-Text API. This API facilitates direct feature extraction and can be integrated with various types of applications, including web, mobile, and desktop platforms. The API is publicly available on GitHub [31].

2) Mobile Application: To utilize the API, we have developed a mobile application that records the audio, allowing users to extract features directly in a structured format. The prototype of the mobile application is illustrated in Figure 2.

TABLE I
NUMBER OF SUBJECTS AND H&Y, UPDRS II-5, UPDRS III-18 SCORES

| | Severity | Number of Subjects |
|---|---|---|
| H&Y | 2 | 8 |
| | 3 | 6 |
| | 4 | 1 |
| UPDRS II-5 | 0 | 7 |
| | 1 | 6 |
| | 2 | 2 |
| UPDRS III-18 | 0 | 6 |
| | 1 | 6 |
| | 2 | 3 |

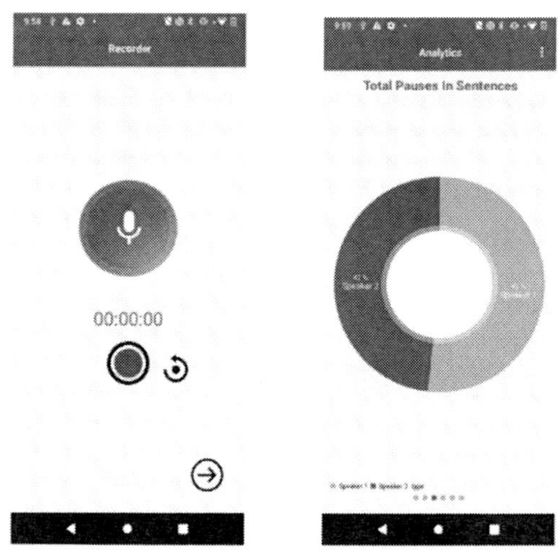


Figure 2. User Interface of Mobile Application: a) (left) recorder, b) (right) demonstration of total speaking time between two speakers.

3) Patient Representation Extraction: The extracted features were then utilized to derive the subjects' hidden representations. We employed an AE approach to learn these representations.

4) Evaluation: We used the hidden representations to distinguish between PD and controls by applying classification algorithms such as KNN, SVM, XGBoost and RF. Moreover, we demonstrated that these hidden representations of the subjects are distinguishable based on their Unified Parkinson's Disease Rating Scale (UPDRS) scores.

### B. Dataset

We used a publicly available dataset from King's College London (KCL) [32], with audio recordings made using a Motorola Moto G4 smartphone. This dataset includes assessments such as the Hoehn and Yahr (H&Y) scale, UPDRS III-18, and UPDRS-II-5 scores. H&Y indicates PD progression stages: '2' for unimpaired bilateral movements, '3' for postural impairment, and '4' for needing assistance with regular activity [33]. The UPDRS III-18 score, assessing the motor examination of speech, ranges from '0' (normal), '1' (slight loss of dictation), to '2' (moderately impaired) [34][35]. The UPDRS II-5 score evaluates daily speech, where '0' is normal and higher scores indicate increasing severity. The study comprised 15 PD subjects and 21 control subjects. The distribution of patients across H&Y and UPDRS is detailed in Table I.

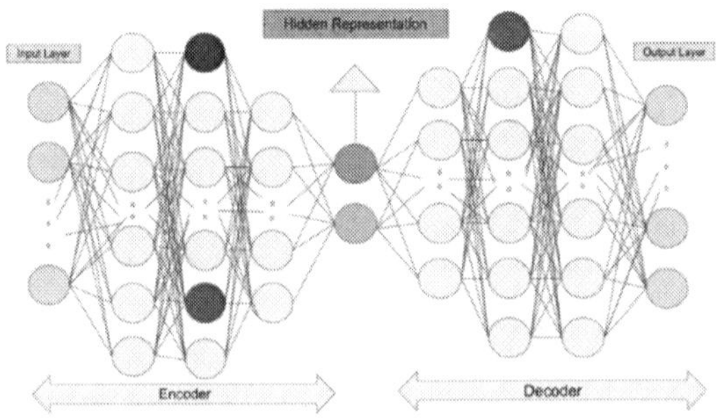


Figure 3. 9-layered stacked autoencoder architecture. Encoder and decoder both consist of 3 fully connected layers and one drop out layer.

### C. Feature Extraction

Using our API [31], we extracted a total of 42 features as listed below, focusing on speaker segmentation and conversation dynamics.

- Total time of each speaker with all the pauses
- Total pause time during a conversation
- Gap between the turns
- Continuous repeating word
- Percentage of speaking time in the conversation
- Total turns
- Total unique words
- Average word length
- Percentage of total first half speaking time
- Percentage of total last half speaking time
- Total conversation duration between 2 people

Additionally, we quantified each Part Of Speech (POS) in the conversations [36] utilizing the Python Natural Language ToolKit (NLTK) [37].

### D. Autoencoder Architecture

We proposed a 9-layer stacked AE, depicted in Figure 3, featuring a fully connected architecture. The AE comprises three main parts: an encoder, a decoder, and a middle code representing the hidden layer. The encoder consists of an input layer, followed by a second layer with 100 neurons, a dropout layer set at 0.1, and a third layer with 30 neurons. The decoder mirrors the encoder in reverse order, aiming to reconstruct the original input at the output layer. The middle layer is fixed at 2 neurons, serving as the hidden representation.

For each subject, we extracted two-dimensional hidden representations. To standardize these embeddings, we normalized the PD subjects' data relative to the control group by calculating the mean and standard deviation for each dimension among the controls and adjusting the PD values using the following equation:

$$AE\_scaled_{PD} = (Mean(AE_{control})) - (AE_{PD}/std(AE_{control}))$$

### E. Hyperparameter Tuning

To optimize the hyperparameters of the AE, we employed a grid search strategy. We tested various neuron counts for the first dense layer (64, 96, 100, 128) and the third dense layer (8, 10, 16, 20, 24, 30, 32) in the encoder. In the decoder, the layers were structured with the same number of neurons as the encoder but in reverse order. Rectified Linear Unit (ReLU) and tanh were applied in hidden layers as activation functions and different learning rates (0.1, 0.01 and 0.001) were explored. After the grid search, we chose the hyperparameters that captured the most variance, as listed in Table II.

TABLE II
HYPERPARAMETERS OF AE

| | Layers | Units | Activation Functions |
|---|---|---|---|
| Encoder | Input Layer | 42 | |
| | Dense Layer | 100 | tanh |
| | Dropout | 0.1 | |
| | Dense Layer | 30 | tanh |
| Code | Dense | 2 | tanh |
| Decoder | Dense Layer | 30 | tanh |
| | Dropout | 0.1 | |
| | Dense Layer | 100 | tanh |
| | Output Layer | 42 | linear |

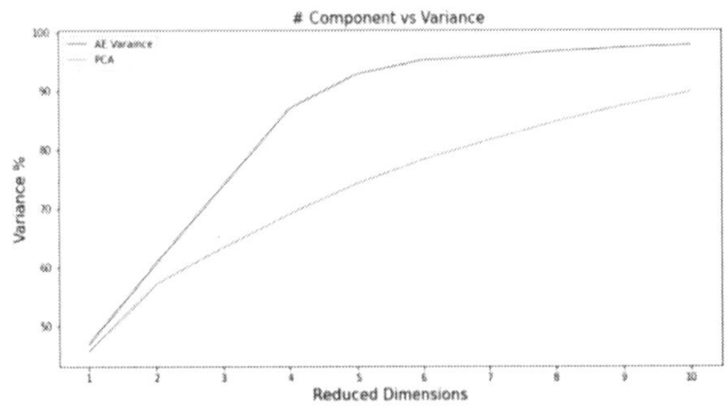


Figure 4. Reduced dimension vs variance. AE captures more variance than PCA.

### F. Classification Models

For comparisons, we employed traditional classifiers including KNN, SVM, RF and XGBoost algorithms. These algorithms were fed with two sets of AE hidden representations, raw features, and two-dimensional PCA features to classify PD versus control subjects. All data were scaled using min-max scaling. Hyperparameters were selected via a grid search conducted on the raw 42 features, and these parameters were then applied consistently across all models. Model performance was assessed using 5-fold cross-validation, and we reported the average validation results from the 5 folds.

## III. RESULTS AND DISCUSSION

Figure 4 illustrates the variance, explained by the trained AE compared to PCA, clearly showing that AE captures more variance. For downstream analysis, we took our 2 hidden representations generated by the encoder and applied normalization to the PD embeddings based on the control subject embeddings. Then, we classified these embeddings using KNN, XGBoost, RF, and SVM algorithms. Figure 5 compares the accuracy of three data modalities across four classifiers and Figure 6 shows the 2D embeddings with UPDRS-III-18 scores, while the mean macro precision, recall, and F1-scores from a 5-fold validation are reported in Table III.

The aim of this study is to enhance insights into conversations via our mobile application, particularly for individuals with speech impairments who need monitoring of their social engagement. The feature extraction API leverages the Google speech recognition system to extract 42 features [38], which are readily accessible through our mobile application designed for two-speaker settings, given the dataset constraints.

Utilizing these features, we developed an AE that captures more variance than PCA. The hidden representations extracted from the AE were used in downstream analysis, showing a significant improvement in classification accuracy for PD versus control subjects, with KNN achieving a 90% macro precision. XGBoost and RF also

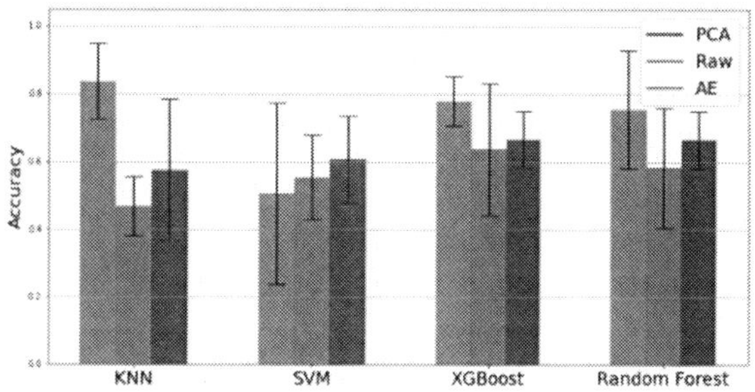


Figure 5. Mean accuracy comparisons using PCA (blue), AE (green), and raw (red) features.

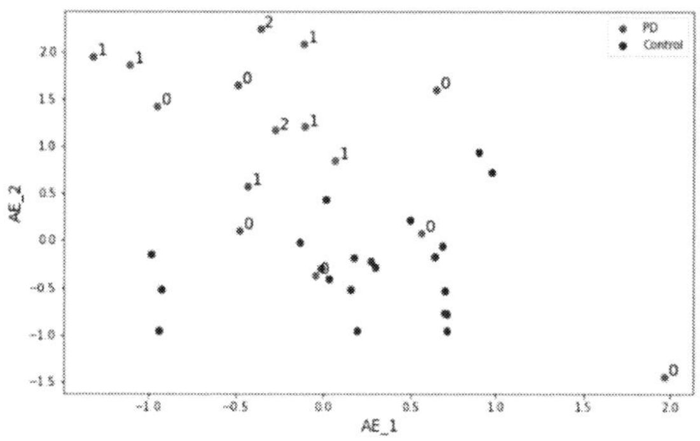


Figure 6. Autoencoder hidden representation visualization for control (blue) and PD (green) with UPDRS-II-5 rating.

showed notable improvements using AE features, although raw-SVM outperformed both AE and PCA. This is because KNN, XGBoost, and RF benefit from the noise reduction and complex pattern representation in the transformed space. SVM, however, performed better with the original features, likely due to its effectiveness in using simpler, direct features for maximizing class separation.

Furthermore, in Figure 6, the hidden representations of three patients, who are close to controls and in the early stages according to the H&Y, UPDRS II-5, and UPDRS III-18 scales, were misclassified as controls by the classification algorithms. This misclassification suggests that their communication skills may not be significantly affected. Additionally, the existing rating systems vary from one to another. For instance, there are seven subjects at stage 0 according to UPDR II-5, but only six according to UPDRS III-18. To address these inconsistencies, we propose using an AE to develop a unified scaling system for assessing subjects. A limitation of this study is the small sample size and the use of control subjects' hidden representations as a baseline for scaling. With a larger dataset, we could directly establish a scale that more accurately evaluates communication skills. This pilot study demonstrates the utility of embeddings in distinguishing PD from control subjects, with less severe PD cases tending to cluster closer to controls, as depicted in Figure 6.

TABLE III
MEAN MACRO PRECISION, RECALL AND F1 SCORE COMPARISONS

| Model | Features | Macro precision | Macro recall | Macro F1 |
|---|---|---|---|---|
| KNN | AE | 0.90 | 0.80 | 0.80 |
| | RAW | 0.26 | 0.40 | 0.32 |
| | PCA | 0.45 | 0.53 | 0.47 |
| XGBoost | AE | 0.81 | 0.76 | 0.75 |
| | RAW | 0.66 | 0.62 | 0.62 |
| | PCA | 0.67 | 0.61 | 0.57 |
| Random Forest | AE | 0.76 | 0.73 | 0.72 |
| | RAW | 0.62 | 0.55 | 0.54 |
| | PCA | 0.62 | 0.60 | 0.54 |
| SVM | AE | 0.48 | 0.50 | 0.48 |
| | RAW | 0.57 | 0.54 | 0.52 |
| | PCA | 0.56 | 0.55 | 0.50 |

Although several previous studies have tried to distinguish between PD and control subjects using deep learning algorithms to assist clinicians in accurately diagnosing PD, many of them are not designed for easy integration with new environments or for extracting features from different datasets. Our proposal includes an application and a versatile API that can be integrated into any platform, offering clinicians deeper insights into patient conversations and social interactions. In the future, with greater data availability, we can develop more accurate models for predicting the severity of communication impairments.

## IV. CONCLUSION

In this paper, we have developed a user-friendly application that extracts features from interactive conversations. We then introduced an AE-based model that generates reduced representations of individuals' social engagement features. Through these hidden representations, the model enabled classification between PD and control groups using embeddings employed by KNN, SVM, RF, and XGBoost classifiers. Our findings showed that these embeddings capture diverse data patterns and effectively distinguish PD patients from control subjects, demonstrating that these representations can encapsulate the full range of an individual's social participation. This highlights the AE model's value in creating meaningful representations for assessing social engagement.

Moving forward, enhancing this model with additional conversational and multimodal features, such as gesture or facial expression data, could further improve its accuracy and adaptability to real-world scenarios. Such expansions would support clinicians in monitoring PD symptoms more effectively and could potentially enable early interventions based on real-time social interaction insights. This study highlights the potential of embedding-based methods in healthcare, offering a pathway toward practical, non-invasive tools that can aid in diagnosing and managing communication disorders through assessment of social engagement patterns.

## ACKNOWLEDGEMENT

The authors would like to thank Thasina Tabashum and Dr. Mark Albert for their help and support in making this work possible. Their expertise and encouragement have been instrumental in enabling us to achieve our research goals.

## REFERENCES

[1] R. J. Ruben, "Redefining the survival of the fittest: Communication disorders in the 21st century," *The Laryngoscope*, vol. 110, no. 2, pp. 241–241, 2000.

[2] R. Body, M. Perkins, and S. McDonald, "Pragmatics, cognition, and communication in traumatic brain injury," in *Communication Disorders Following Traumatic Brain Injury*. Psychology Press, 2016, pp. 81–112.

[3] C. K. Thompson, "The American Speech-Language-Hearing Association Functional Assessment of Communication Skills ASHA FACS: Test description and pilot data," in *A Paper Presented at the Clinical Aphasiology Conference, Sunriver, OR, 1995*, 1995, pp. 75–78.

[4] World Health Organization, *International Classification of Functioning, Disability, and Health: Children & Youth Version: ICF-CY.* World Health Organization, 2007.

[5] R. O'Halloran, M. Carragher, and A. Foster, "The consequences of the consequences: The impact of the environment on people with aphasia over time," *Topics in language disorders*, vol. 37, no. 1, pp. 85–100, 2017.

[6] H. L. Flowers, S. A. Skoretz, F. L. Silver, E. Rochon, J. Fang, C. Flamand-Roze, and R. Martino, "Poststroke aphasia frequency, recovery, and outcomes: A systematic review and meta-analysis," *Archives of Physical Medicine and Rehabilitation*, vol. 97, no. 12, pp. 2188–2201, 2016.

[7] K. Brown, B. Davidson, L. E. Worrall, and T. Howe, ""Making a Good Time": The role of Friendship in Living Successfully with Aphasia," *International Journal of Speech-Language Pathology*, vol. 15, no. 2, pp. 165–175, 2013.

[8] K. Hilari, S. Northcott, P. Roy, J. Marshall, R. D. Wiggins, J. Chataway, and D. Ames, "Psychological distress after stroke and aphasia: The first six months," *Clinical Rehabilitation*, vol. 24, no. 2, pp. 181–190, 2010.

[9] D. R. Paul, C. Frattali, A. L. Holland, C. K. Thompson, C. J. Caperton, and S. C. Slater, *Quality of Communication Life Scale: Manual.* American Speech-Language Hearing Association, 2004.

[10] P. Sukanya and B. S. Rao, "Parkinson's disease classification based on stacked denoising autoencoder," *Computer Science*, vol. 24, no. 4, 2023.

[11] T. Xiao, T. Tabashum, B. Metwally, M. Albert, A. Du, R. Jebamalaidass, M. Leal, and E. Oliviera, "Conversation moderator: A mobile app for tracking individual speaking in group conversations," in *2020 IEEE 14th International Conference on Semantic Computing (ICSC).* IEEE, 2020, pp. 430–433.

[12] C. G. Goetz *et al.*, "Testing objective measures of motor impairment in early parkinson's disease: Feasibility study of an at-home testing device," *Movement Disorders*, vol. 24, no. 4, pp. 551–556, 2009.

[13] J. Rusz, J. Hlavnička, T. Tykalová, M. Novotný, P. Dušek, K. Šonka, and E. Růžička, "Smartphone Allows Capture of Speech Abnormalities Associated with High Risk of Developing Parkinson's Disease," *IEEE Transactions on Neural Systems and Rehabilitation Engineering*, vol. 26, no. 8, pp. 1495–1507, 2018.

[14] A. Tsanas, M. A. Little, and L. O. Ramig, "Remote assessment of parkinson's disease symptom severity using the simulated cellular mobile telephone network," *IEEE Access*, vol. 9, pp. 11 024–11 036, 2021.

[15] P. McNamara and R. Durso, "Pragmatic Communication Skills in Patients with Parkinson's Disease," *Brain and Language*, vol. 84, no. 3, pp. 414–423, 2003.

[16] G. A. Robinson, "Primary Progressive Dynamic Aphasia and Parkinsonism: Generation, Selection and Sequencing Deficits," *Neuropsychologia*, vol. 51, no. 13, pp. 2534–2547, 2013.

[17] J. Dushanova, *Diagnostics and Rehabilitation of Parkinson's Disease.* BoD–Books on Demand, 2011.

[18] A. Yeung *et al.*, "Correlating natural language processing and automated speech analysis with clinician assessment to quantify speech-language changes in mild cognitive impairment and alzheimer's dementia," *Alzheimer's Research & Therapy*, vol. 13, no. 1, p. 109, 2021.

[19] J. R. Orozco-Arroyave, *Analysis of Speech of People with Parkinson's Disease.* Logos Verlag Berlin GmbH, 2016, vol. 41.

[20] L. Berus, S. Klancnik, M. Brezocnik, and M. Ficko, "Classifying parkinson's disease based on acoustic measures using artificial neural networks," *Sensors*, vol. 19, no. 1, p. 16, 2018.

[21] A. Tsanas, M. A. Little, C. Fox, and L. O. Ramig, "Objective automatic assessment of rehabilitative speech treatment in parkinson's disease," *IEEE Transactions on Neural Systems and Rehabilitation Engineering*, vol. 22, no. 1, pp. 181–190, 2013.

[22] K. Pearson, "LIII. On Lines and Planes of Closest Fit to Systems of Points in Space," *The London, Edinburgh, and Dublin philosophical magazine and journal of science*, vol. 2, no. 11, pp. 559–572, 1901.

[23] D. E. Rumelhart, G. E. Hinton, and R. J. Williams, "Learning representations by back-propagating errors," *Nature*, vol. 323, no. 6088, pp. 533–536, 1986.

[24] S. Nandigam, T. Tabashum, and T. Xiao, "Autoencoders and embeddings: How unsupervised structural learning enables fast and efficient goal-directed learning," in *Bridging Human Intelligence and Artificial Intelligence.* Springer, 2022, pp. 47–63.

[25] S.-I. Ng and T. Lee, "Automatic Detection of Phonological Errors in child Speech Using Siamese Recurrent Autoencoder," *arXiv preprint arXiv:2008.03193*, 2020.

[26] A. Caliskan, H. Badem, A. Basturk, and M. Yuksel, "Diagnosis of the parkinson disease by using deep neural network classifier," *IU-Journal of Electrical & Electronics Engineering*, vol. 17, no. 2, pp. 3311–3318, 2017.

[27] H. Gunduz, "Deep learning-based parkinson's disease classification using vocal feature sets," *IEEE access*, vol. 7, pp. 115 540–115 551, 2019.

[28] N. Cristianini and J. Shawe-Taylor, *An Introduction to Support Vector Machines and Other Kernel-Based Learning Methods.* Cambridge University Press, 2000.

[29] T. Chen and C. Guestrin, "XGBoost: A Scalable Tree Boosting System," in *Proceedings of the 22nd ACM SIGKDD International Conference on Knowledge Discovery and Data Mining*, ser. KDD '16, vol. 11. ACM, Aug. 2016, p. 785–794. [Online]. Available: http://dx.doi.org/10.1145/2939672.2939785

[30] M. Hoq, M. N. Uddin, and S.-B. Park, "Vocal feature extraction-based artificial intelligent model for parkinson's disease detection," *Diagnostics*, vol. 11, no. 6, p. 1076, 2021.

[31] T. Xiao, P. Guha, T. Thasina, and A. Tajuddin (2024). Conversation Feature Extraction Tool. Deep 6 Lab. https://github.com/Deep6Lab/Conversation-Feature-Extraction-Tool. Accessed: 2024-10-31.

[32] H. Jaeger, D. Trivedi, and M. Stadtschnitzer, "Mobile device voice recordings at king's college london (mdvr-kcl) from both early and advanced parkinson's disease patients and healthy controls," *Zenodo*, 2019.

[33] M. M. Hoehn and M. D. Yahr, "Parkinsonism: Onset, progression, and mortality," *Neurology*, vol. 17, no. 5, pp. 427–427, 1967.

[34] Parkinson's Disease Research, Education and Clinical Centers, U.S. Department of Veteran Affairs. https://www.parkinsons.va.gov/resources/UPDRS.asp. Accessed: 2024-10-31.

[35] N. Diaz and J. M. Bronstein, "Parkinson's disease research education and clinical centers (PADRECC): background and overview," *NeuroRehabilitation*, vol. 20, no. 3, pp. 153–160, 2005.

[36] S. Bird, E. Klein, and E. Loper, *Natural Language Processing with Python*, 01 2009.

[37] B. Steven, E. Loper, and E. Klein (2009). Natural Language Processing with Python. O'Reilly Media Inc. https://www.nltk.org/. Accessed: 2024-10-31.

[38] F. Filippidou and L. Moussiades, "A Benchmarking of IBM, Google and Wit Automatic Speech Recognition Systems," in *16th IFIP International Conference on Artificial Intelligence Applications and Innovations (AIAI)*, ser. IFIP Advances in Information and Communication Technology, I. Maglogiannis, L. Iliadis, and E. Pimenidis, Eds., vol. AICT-583, no. Part I. Neos Marmaras, Greece: Springer International Publishing, Jun. 2020, pp. 73–82, part 1: Classification. [Online]. Available: https://inria.hal.science/hal-04050587

# Does Complexity Pay Off? Applying Advanced Algorithms to Depression Detection on the GLOBEM Dataset

Sebastian Cavada, Alvaro Berobide, Yevheniia Kryklyvets
Mohamed Bin Zayed University of Artificial Intelligence, Abu Dhabi, UAE
e-mail: {name.surname}@mbzuai.ac.ae

***Abstract*—This manuscript evaluates the performance of state-of-the-art time series analysis algorithms for depression detection on the Generalization of LOngitudinal BEhavior Modeling (GLOBEM) dataset. We assess *Time-Series Mixer (TSMixer), Cross-former, Gated Recurrent Unit (GRU), Convolutional Neural Network with Long Short-Term Memory (CNN_LSTM)* and introduce a novel self-developed algorithm with the goal of increasing accuracy over the original *Reorder*. While these models demonstrate robust out-of-domain generalization, they fail to surpass the accuracy of the baseline *Reorder* algorithm, which was specifically developed for in-domain analysis by the GLOBEM team. Our findings reveal consistently low performance across all models, suggesting limitations inherent in the dataset rather than the algorithms themselves. We hypothesize that the dataset's absence of critical variables and insufficient granularity likely limits model convergence. This hypothesis is supported by similar studies that achieved higher accuracy using more frequent data points with similar architecture approaches. Based on these insights, we suggest that future studies might benefit from incorporating more granular sensor measurements and more sophisticated data types, such as, but not limited to, Heart Rate Variability (HRV).**

***Keywords-Depression Detection; Time-Series Analysis; Deep Learning; Domain Generalization; Mental Health.***

## I. INTRODUCTION

It is estimated that 3.8% of the global population suffers from clinical depression condition. This mood disorder affects over 280 million people, ranking it among the leading causes of disability [1]. Despite its prevalence, this condition remains challenging to diagnose and treat effectively, often due to delayed detection. Traditional diagnostic methods, relying on subjective assessments, can miss early warning signs. This underscores the need for objective, data-driven approaches to enable earlier and more accurate diagnosis [2] by building applications that will allow for self-monitoring and alerting when professional assistance is required.

In particular, recent advancements in wearable hardware have enabled continuous monitoring of human physiological data, including heart rate, oxygen levels, and movement patterns. This wave of technology sparked interest in the deep learning community to leverage this temporal information to develop automated methods for depression detection [3][4][5]. Despite these innovations, the efficacy of such approaches remains limited, with results often being minimally informative and thus remaining a subject of ongoing research and debate [3].

To the best of our knowledge, this study represents the first evaluation of state-of-the-art time series analysis algorithms for depression detection tasks using the GLOBEM dataset [3]. We examine various advanced models, including *TSMixer* [6], *Crossformer*[7], *GRU* [8], *CNN_LSTM* [9]. Additionally, we introduce a novel algorithm that enhances the baseline *Reorder* [3] with LSTM capabilities. The aim for the new model is to improve the current *Reorder* algorithm. By adapting all these models, we aim to give a snapshot of the current state of depression detection algorithms and emphasize a critical finding: the key to improvement may lie in better data rather than more complex algorithms.

The remainder of the paper is organized as follows: In Section II, we present a review of related work in the field of automated depression detection. Section III details the methodology of our study, including the dataset used and the algorithms evaluated. Section IV presents our results. In Section V, we discuss the implications of our results and the limitations of current approaches. Section VI offers the conclusion and directions for future research.

## II. RELATED WORK

Our research focuses on the application of Artificial Intelligence (AI) to address critical health issues like depression, leveraging multi-year longitudinal data. The GLOBEM dataset [3] stands out as a pioneering dataset culled from a comprehensive multi-year data collection study, capturing a broad spectrum of data from 497 unique participants, totaling 705 person-years.

In the field of Time-Series Forecasting (TSF), transformers have revolutionized sequence modeling with their unparalleled performance across domains [10]. However, their application in TSF, especially for long-term forecasting, has yielded mixed results. Some studies have highlighted limitations [11], while others suggest that transformers may still hold untapped potential in this area [12]. The all-Multi-Layer Perceptron (MLP) architecture, initially conceived for Computer Vision [13], has been repurposed for TSF through the *TSMixer* work [14], enabling the handling of multivariate data and highlighting the adaptability to large datasets and complex real-world scenarios Recurrent Neural Networks (RNNs) [15] and their variants [8] have long been the standard approach for time series forecasting. Their ability to handle sequential data has made them particularly useful for multivariate time series prediction over many years.

Domain generalization in time-series prediction encompasses various related works aimed at developing models capable of performing well across different domains without the need for domain-specific training data [16][17][18]. These methodologies address the challenge of domain shift, enabling models to generalize effectively across diverse domains.

## III. METHODS

This section provides a comprehensive overview of our methodology, covering four key areas: Dataset description, algorithms, experimental setup, and implementation details.

### A. Dataset Description

The GLOBEM dataset spans four years and includes data from 705 person-years [19][20]. It consists of two primary data types: survey data and passive mobile sensing data.

Survey data, collected periodically throughout the study, includes metrics from the Beck Depression Inventory-II (BDI-II) and the Patient Health Questionnaire-4 (PHQ-4), which serve as ground truth for depression and anxiety. This data is critical for the binary classification of mental health states (if the pathology is present or not), providing insights into the severity of symptoms across a diverse population.

Passive mobile sensing data gathered via a dedicated app on iOS and Android devices and Fitbit wearable tracks location, phone usage, physical activity, and other behaviors in real-time. This extensive data set, with more than 1000 distinct features from phone usage alone (extracted and standardized by the Reproducible Analysis Pipeline for Data Streams Open Source platform [21]), is crucial to analyzing daily routines and behaviors, offering a comprehensive view of the impacts of lifestyles on mental health. Given the high dimensionality of the raw data, a rigorous feature selection and data preparation process was implemented. This process aimed to distill the most impactful insights while managing computational complexity. The final prediction model utilizes a subset of 54 key features selected for their relevance and predictive power. Data is structured in batches, each representing a 4-week (28-day) period. This temporal structure allows for analyzing both short-term fluctuations and longer-term behavioral patterns, enhancing the dataset's utility for depression detection tasks.

### B. Algorithms

*1) Reorder - the baseline algorithm:* The *Reorder* algorithm is a multi-task learning model that uses the continuity of behavior trajectories to enhance domain generalization in behavioral models; the details are shown in Figure 1. Its primary goal is maintaining time continuity while addressing a principal classification task. It optimizes two distinct losses simultaneously: the binary cross-entropy loss, based on the ground truth, and a second loss from a self-supervised task. This task involves predicting the position of segments randomly shuffled in a subset of all possible permutations. This self-supervised task act as a regularizer, encouraging the network to understand the temporal dimension and improve generalization to the main task.

*2) TSMixer - All-MLP Architecture:* The *TSMixer* architecture, part of the ALL-MLP family, is chosen for its ability to efficiently handle multivariate time-series data through MLPs. This model simplifies complex pattern recognition across time

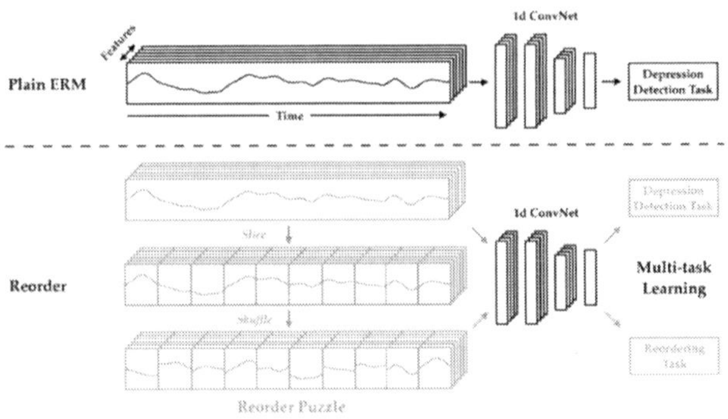


Figure 1. The *Reorder* architecture, image from the original paper [3].

and feature dimensions, making it suitable for the computational demands of depression diagnosis prediction [6].

*3) Crossformer - Transformer Based Model:* We utilize the *Crossformer* architecture due to its advanced capacity for handling long-term dependencies and high-dimensional data. Its hierarchical integration of features allows for a nuanced understanding of time-series patterns, crucial for accurate predictions in time-series data [7].

*4) Utilization of Models from the RNN Family:* RNNs, including *GRU* and *LSTM* variants, are employed for their unique ability to maintain a memory of past meaningful information, enabling effective modeling of time-dependent data. This characteristic is particularly beneficial for tracking the progression of depressive symptoms over time [8].

*5) Reorder + CNN_LSTM:* Our top-performing model, as shown in Table I, is a novel algorithm that we have named *Reorder + CNN_LSTM*. This algorithm combines the strengths of the original Reorder model [3] with the capabilities of a CNN_LSTM architecture. The LSTM module was added particularly to capture long-term dependencies in the sequence, enhancing the model's ability to recognize patterns over extended periods.

This hybrid approach used three times as many parameters as the original *Reorder* but allowed us to leverage the benefits of each individual component:

- Reorder: Effective temporal data handling
- CNN: Spatial feature extraction
- LSTM: Long-term dependency learning and improved generalization

The 32,138 parameters result from merging the different models: Reorder and CNN_LSTM. Some parameters are shared among various modules, such as the initial and final layers, therefore the total number of parameters doesn't exactly add up to the individual number of parameters of each model.

We report parameter count as a key indicator of model complexity, especially relevant in resource-constrained environments.

### C. Experimental Setup

In this research, we adhered to the experimental framework presented in the GLOBEM paper [3] to ensure the comparability

TABLE I
ALL RESULTS ARE IN DESCENDING ORDER, OUR METHODS IN DIFFERENT COLORS, RESULTS ARE IN BALANCED ACCURACY. THE STANDARD DEVIATION IS CALCULATED ON THE NUMBER OF RUNS BETWEEN THE DATASETS. **The number of parameters takes into account only trainable parameters - The comma is used to separate thousands, while the point is used for decimals.*

| Model | Number of Parameters* | Results | | |
|---|---|---|---|---|
| | | Single Dataset | Leave one out | Pre/Post Covid |
| Reorder + CNN-LSTM | 32,138 | 0.629 ± 0.045 | 0.542 ± 0.009 | 0.530 ± 0.001 |
| **Reorder** | **10,162** | **0.626±0.063** | **0.548±0.030** | **0.513±0.009** |
| **CNN-LSTM** | **24,378** | **0.601±0.026** | **0.513±0.009** | **0.507±0.004** |
| GRU | 62,226 | 0.591±0.034 | 0.516±0.011 | 0.502±0.001 |
| Crossformer | 131,527 | 0.590±0.001 | 0.503±0.003 | 0.516±0.002 |
| ERM-Transformer | 12,354 | 0.584±0.013 | 0.509±0.008 | 0.512±0.016 |
| IRM | 2,698 | 0.573±0.016 | 0.506±0.006 | 0.499±0.000 |
| ERM-1dCNN | 2,698 | 0.568±0.006 | 0.510±0.008 | 0.514±0.006 |
| ERM-Mixup | 2,698 | 0.568±0.006 | 0.501±0.008 | 0.507±0.004 |
| ERM-LSTM | 22,186 | 0.565±0.019 | 0.512±0.006 | 0.512±0.003 |
| TSMixer | 43,429 | 0.543±0.035 | 0.521±0.006 | 0.499±0.000 |
| CSD-D | 2,839 | 0.562±0.022 | 0.521±0.002 | 0.512±0.006 |
| Siamese Network | 2,664 | 0.545±0.025 | 0.509±0.010 | 0.515±0.002 |
| CSD-P | 2,875 | 0.542±0.010 | 0.511±0.006 | 0.516±0.000 |
| ERM-2dCNN | 12,994 | 0.533±0.013 | 0.510±0.006 | 0.504±0.006 |
| DANN-D | 3,281 | 0.526±0.016 | 0.514±0.004 | 0.514±0.000 |
| MLDG-D | 2,698 | 0.522±0.013 | 0.511±0.006 | 0.495±0.004 |
| MLDG-P | 2,698 | 0.508±0.011 | 0.510±0.003 | 0.500±0.003 |
| MASF-D | 2,970 | 0.505±0.006 | 0.505±0.001 | 0.504±0.007 |
| DANN-P | 3,578 | 0.502±0.002 | 0.500±0.000 | 0.500±0.000 |
| MASF-P | 2,970 | 0.495±0.007 | 0.505±0.004 | 0.509±0.011 |

of our results. Our experiments were designed to evaluate the performance of algorithms in three distinct scenarios:

1) **Single Dataset** This method divides the data for each participant within a dataset, using the first 80% for training and the remaining 20% for testing. This setup assesses the model's predictive capability using past data to forecast future outcomes.
2) **Leave-One-Dataset-Out:** In this cross-dataset approach, the model is trained on three datasets and tested on the fourth. This configuration evaluates the model's generalizability across different datasets.
3) **Pre/Post-COVID Analysis:** This setup aims to discern the impact of the COVID-19 pandemic on model performance. It involves training on datasets INS-1 (Data Set year 1) and INS-2 (pre-COVID) and testing on INS-3 and INS-4 (post-COVID), with a subsequent reversal of training and testing datasets to examine the effects comprehensively. The different particularities of the dataset are explained in the following "Dataset Description" section.

### D. Implementation details

All computational experiments were conducted on a high-performance workstation equipped with a GPU 4090 and 44GB of RAM, using TensorFlow and Keras for model implementation. We adopted the Adam optimizer with an initial learning rate of 0.001, adjusted by cosine annealing with a decay rate of 0.95 and a step size of 20. The models were trained for up to 200 epochs with early stopping based on the best validation loss, allowing a minor degree of data leakage, as noted in the original paper. Consistent with established protocols to ensure a fair comparison with previous studies, we used balanced accuracy as our main evaluation metric [3]. This metric, calculated as $\frac{1}{2}\left(Sensitivity + Specificity\right)$, where $Sensitivity = \frac{TruePositive}{TruePositive+FalseNegative}$ and $Specificity = \frac{TrueNegative}{TrueNegative+FalsePositive}$, is particularly effective in contexts with class imbalances. Using balanced accuracy allows us to accurately assess and compare the performance of our proposed approach against existing methods, providing a robust measure of effectiveness across diverse models and datasets.

## IV. RESULTS

Table I reports the balanced accuracy for all methods, ordered by performance on the single dataset. Our experiments are highlighted in different colors, while results from the original paper are in black.

Three of the four State-Of-The-Art (SOTA) models implemented in this study outperform nearly all the models discussed in the original paper, except for their top model, *Reorder*. The gap between *Reorder* and the best-adapted model is only 2%, with more recent models lagging behind despite having at least twice as many parameters. This decreased return in performance is observed across all three tasks, with Crossformers being one exception noted below.

Our novel model, *Reorder + CNN_LSTM*, achieved the highest performance in Table I. It showed a slight increase in accuracy (0.3%) on the single dataset; conversely, it improved accuracy on the Pre/Post Covid dataset by a non-trivial 2%. It also showed negligible lowered performance in the leave-one-out dataset. These improvements come with a cost of three times more parameters, as mentioned before.

The Pre/Post COVID dataset proved to be the most challenging task, likely due to lifestyle disruptions in individuals. Higher accuracy on this dataset indicates better model robustness to strong shifts in the test domain. Notably, the *Crossformer* model surpassed the baseline in this task by nearly 1%, but at the cost of having almost 10 times as many parameters.

The fourth SOTA model analyzed, *TSMixer*, significantly underperformed in all tasks, lagging behind both other SOTA models and older deep learning approaches despite requiring a substantial increase in the number of parameters.

## V. Discussion

Our comprehensive evaluation of SOTA algorithms and original deep learning methods for depression detection using wearable data has revealed several important insights. Across all methods, we observed consistently low accuracies, a finding that aligns with Xu et al. [3], who noted that "Current cross-dataset generalizability algorithms are still far from satisfactory for real-life deployment." This persistent challenge suggests that despite the variety of algorithms employed, the data itself might lack sufficiently informative values for reliable depression detection.

The limitations of the dataset, as acknowledged by its original authors, are particularly noteworthy. The absence of certain sensor signals, such as Heart Rate Variability and Saturation of peripheral Oxygen (SpO2) measures, may be critical missing variables needed to increase accuracy and more reliably detect depression [3][22]. This observation is further supported by research on more granular data, such as minute-per-minute Heart Rate Variability, which has achieved higher accuracy rates of around 71% in similar settings [22].

Our results also indicate that increased model complexity does not necessarily translate to improved performance. The novel *Reorder + CNN_LSTM* algorithm demonstrated only considerable improvements over the original *Reorder* in one out of three tasks, raising questions about the cost-benefit ratio of increased model complexity. Similarly, the poor performance of the TSMixer model, despite its increased parameter count, suggests that its linear nature may not adequately capture the intricacies of this particular time series multivariate distribution.

## VI. Conclusion and Future Work

In conclusion, our research adapted new state-of-the-art time series analysis algorithms, specifically *TSMixer*, *Crossformer*, *GRU*, and *CNN_LSTM* for depression detection on the GLOBEM dataset. While these algorithms exhibited robust out-of-domain generalizability with balanced accuracy on par with specialized architectures, they did not surpass the baseline *Reorder*. We also introduced a novel variant of the *Reorder* algorithm, which improved performance, especially on the Covid cross-dataset. Nevertheless, the baseline *Reorder* still maintains superior computational Pareto efficiency, offering the best accuracy-to-parameter ratio.

Our findings indicate that larger, more complex models perform no better than their simpler counterparts in this specific task, similar to results in other studies where *simpler traditional methods* like tree-based models outperformed complex deep models on tabular data [23]. Furthermore, the results remain close to a non-informative baseline, and we suggest that the current dataset may have insufficient variables for reliable depression detection. Future studies might benefit from incorporating both more granular measurements as well as additional data types such as HRV and SpO2. Furthermore, new sensors, such as the electrocardiogram (ECG) from the Apple Watch, will become available as new devices are released on the market enabling new research.

## References


[1] World Health Organization, "Depressive disorder (depression)", *World Health Organization: WHO*, Mar. 2023.

[2] S. Saeb *et al.*, "Mobile Phone Sensor Correlates of Depressive Symptom Severity in Daily-Life Behavior: An Exploratory Study", eng, *Journal of Medical Internet Research*, vol. 17, no. 7, e4273, Jul. 2015, ISSN: 1438-8871. DOI: 10.2196/jmir.4273.

[3] X. Xu *et al.*, *GLOBEM Dataset: Multi-Year Datasets for Longitudinal Human Behavior Modeling Generalization*, arXiv:2211.02733 [cs], Mar. 2023.

[4] P. Chikersal *et al.*, "Detecting Depression and Predicting its Onset Using Longitudinal Symptoms Captured by Passive Sensing: A Machine Learning Approach With Robust Feature Selection", *ACM Transactions on Computer-Human Interaction*, vol. 28, no. 1, pp. 1–41, Jan. 2021, ISSN: 1073-0516. DOI: 10.1145/3422821.

[5] R. Wang *et al.*, "StudentLife: Assessing mental health, academic performance and behavioral trends of college students using smartphones", en, in *Proceedings of the 2014 ACM International Joint Conference on Pervasive and Ubiquitous Computing*, Seattle Washington: ACM, Sep. 2014, pp. 3–14, ISBN: 978-1-4503-2968-2. DOI: 10.1145/2632048.2632054.

[6] S.-A. Chen *et al.*, *TSMixer: An All-MLP Architecture for Time Series Forecasting*, arXiv:2303.06053 [cs], Sep. 2023.

[7] Y. Zhang and J. Yan, "Crossformer: Transformer utilizing cross-dimension dependency for multivariate time series forecasting", in *The Eleventh International Conference on Learning Representations*, 2023.

[8] Y. Zhang, R. Wu, S. M. Dascalu, and F. C. Harris, "A novel extreme adaptive gru for multivariate time series forecasting", *Scientific Reports*, vol. 14, no. 1, p. 2991, Feb. 2024. DOI: 10.1038/s41598-024-53460-y.

[9] H. Widiputra, A. Mailangkay, and E. Gautama, "Multivariate cnn-lstm model for multiple parallel financial time-series prediction", *Complexity*, vol. 2021, p. 14, Oct. 2021. DOI: 10.1155/2021/9903518.

[10] Q. Wen *et al.*, *Transformers in Time Series: A Survey*, arXiv:2202.07125 [cs, eess, stat], May 2023.

[11] A. Zeng, M. Chen, L. Zhang, and Q. Xu, "Are transformers effective for time series forecasting?", in *Proceedings of the AAAI conference on artificial intelligence*, vol. 37, 2023, pp. 11 121–11 128.

[12] *Yes, Transformers are Effective for Time Series Forecasting (+ Autoformer)*, [Online; accessed 1. May 2024], May 2024.

[13] I. O. Tolstikhin *et al.*, "Mlp-mixer: An all-mlp architecture for vision", *Advances in neural information processing systems*, vol. 34, pp. 24 261–24 272, 2021.

[14] N. Gruver, M. Finzi, S. Qiu, and A. G. Wilson, *Large Language Models Are Zero-Shot Time Series Forecasters*, arXiv:2310.07820 [cs], Oct. 2023. DOI: 10.48550/arXiv.2310.07820.

[15] R. M. Schmidt, *Recurrent neural networks (rnns): A gentle introduction and overview*, 2019. arXiv: 1912.05911 [cs.LG].

[16] E. Tzeng, J. Hoffman, K. Saenko, and T. Darrell, "Adversarial discriminative domain adaptation", in *Proceedings of the IEEE conference on computer vision and pattern recognition*, 2017, pp. 7167–7176.

[17] T. Hospedales, A. Antoniou, P. Micaelli, and A. Storkey, "Meta-learning in neural networks: A survey", *IEEE transactions on pattern analysis and machine intelligence*, vol. 44, no. 9, pp. 5149–5169, 2021.

[18] I. Gulrajani and D. Lopez-Paz, *In search of lost domain generalization*, 2020. arXiv: 2007.01434 [cs.LG].

[19] I. B. Weiner and W. E. Craighead, *The corsini encyclopedia of psychology, volume 4*. John Wiley & Sons, 2010, vol. 4.

[20] J. Stanhope, "Patient Health Questionnaire-4", *Occupational Medicine*, vol. 66, no. 9, pp. 760–761, Dec. 2016, ISSN: 0962-7480. DOI: 10.1093/occmed/kqw165.

[21] *RAPIDS, https://www.rapids.science/1.10*, [Online; accessed 22. Sep. 2024], Sep. 2024.

[22] L. V. Coutts, D. Plans, A. W. Brown, and J. Collomosse, "Deep learning with wearable based heart rate variability for prediction of mental and general health", *Journal of Biomedical Informatics*, vol. 112, p. 103 610, 2020.

[23] L. Grinsztajn, E. Oyallon, and G. Varoquaux, "Why do tree-based models still outperform deep learning on typical tabular data?", *Advances in neural information processing systems*, vol. 35, pp. 507–520, 2022.

# AI-based Estimation of Lower Limb Joint Moments in Stance Phase Using a Single Wearable Inertial Sensor

Kyoko Shibata
Kochi University of Technology
Tosayamada, Kami, Kochi, 782-8502, Japan
e-mail: shibata.kyoko@kochi-tech.ac.jp

Kohei Watanabe
Kochi University of Technology
Tosayamada, Kami, Kochi, 782-8502, Japan
e-mail: kohei.p225@gmail.com

*Abstract*—**Walking is an easy way to exercise that can maintain and improve health. Quantifying the benefits of walking exercise would make health promotion more effective. The purpose of this study is to estimate lower limb joint moments during daily walking in order to support active healthcare by oneself. Using acceleration data acquired from a large number of wearable sensors, it is not possible to estimate joint moments based on kinetic theory alone. Therefore, this study proposes a method for estimating joint moments using deep learning from measured single-axis acceleration data only, considering the ease of measurement. The accuracy of estimation on the three lower limb joint moments in the stance phase is shown and the benefits of the proposed method are discussed.**

***Keywords- Self-healthcare; Gait analysis; Wearable sensing; LSTM.***

## I. INTRODUCTION

One of the quantitative parameters to validate the load of exercise is the lower limb joint moment (joint torque). This is because muscle activity can be estimated from joint moments [1]. Therefore, joint moment is also a parameter used for diagnosis in orthopedic and rehabilitation clinics. In this study, we propose a method to easily obtain joint moments in daily life. If this method can be systematized, we believe that it will contribute to enhancing the effectiveness of exercise by quantitatively and visually confirming the effects of daily health care exercises by oneself. In other words, support for active self-healthcare can be realized. In this study, first, we will estimate the lower limb joint moments during the stance phase of walking exercise in a simplified manner.

The conventional method for obtaining joint moments during gait with high accuracy is generally to calculate them by inverse dynamic theory using statistical values (e.g., mass, center of gravity position, and moment of inertia, of body part) from ground reaction force data and coordinates, acceleration, and angular velocity of body part. The accuracy is high when multiple large installed force plates and an optical motion capture system are used as measurement devices. However, these devices are limited in installation locations and are expensive, so they are limited to use in specialized institutions such as hospitals and rehabilitation facilities, and are not applicable to measurements in daily life. An alternative to these devices is the use of wearable inertial sensors. Kawamura et al. [2] measured body part acceleration with wearable inertial sensors and calculated lower limb joint moments during running from inverse dynamics theory using statistical values. In our previous report [3][4], we also investigated the use of wearable inertial sensors during walking and obtained some results. However, the method to calculate lower limb joint moments from acceleration as in the previous report can only be applied to the single support phase, because the double support phase, which is not present in running but is present in walking, is a statically indeterminate structure. In addition, to ensure high accuracy, the number of sensors must be 15 in order to include the entire body. Furthermore, the accuracy of joint moment estimation falls as the error in the dynamic acceleration measured accumulates as the number of body parts to be considered increases.

Therefore, this study attempts to estimate joint moments using deep learning from the measurement information of wearable inertial sensors. A previous study [5] used machine learning to predict joint moments and even joint angles. This study used multiple parameters simulated from measured data using optical motion capture systems and multiple inertial sensors as input data, and further expanded the data set by data augmentation. In these cases, it is not easy to prepare a large number of sensor systems and intelligent signal processing. Therefore, this proposal uses only one wearable inertial sensor for measurement when the user estimates, even if errors are introduced, and only actual measured data. The creation of a pre-prepared trained deep learning model requires a high degree of accuracy, so force plates and optical motion capture system must be used, but again, only calculated values from actual measured data are used. In addition, only one wearable inertial sensor is used for estimation. In the future, estimation using only users smartphone is a feasible method. This will help the case of effective active self-health care. In this paper, we describe the proposed method and verify the estimation accuracy. Then, we would like to consider whether it is possible to incorporate easy observation of joint moments into daily life.

The rest of this paper is organized as follows. In Section II, we present the proposed estimation method, then explain the walking experiment to acquire deep learning data and its data processing method, and then describe the method for building deep learning models. Section III shows and discusses the estimated results. Finally, Section IV summarizes this paper and describes future work.

## II. Methods

An outline of the proposed method is shown in Figure 1. In the proposed method, three deep learning models are constructed for each joint by learning the relationship between the time series data of single-axis acceleration acquired from wearable sensor and the correct values of three lower limb joint moments in the sagittal plane, respectively. Untrained single-axis acceleration data not used for learning are input to these learned deep learning models, and the estimated values of the joint moments are the output. For simplicity, the acceleration data is the same single-axis time series data for all three joints.

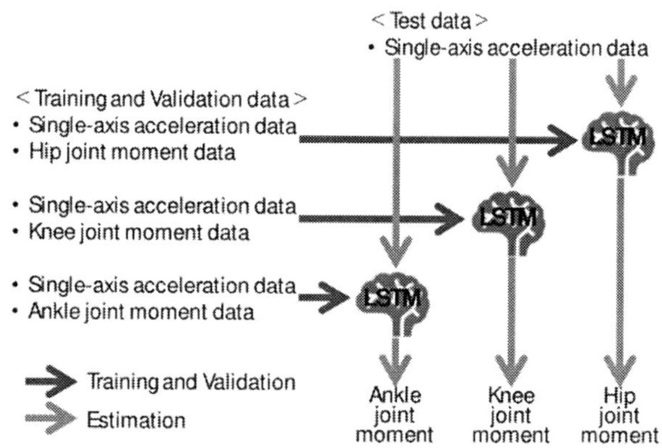


Figure 1. Outline of the proposed method.

Figure 2. Wearing position of inertial sensors.

The experiment for data acquisition is described next.

Two healthy male subjects (age 22±0 years, height 1.66±0.07 [m], weight 74.0±12.7 [kg]) participated in the experiment. This experiment was conducted after obtaining approval from the University's Ethics Review Committee (No. 176) and after explaining the experiment to the subjects and obtaining their consent.

Three force plates (TF-6090, TF-4060: Tech Giken) and an optical motion capture system (MAC 3D System: Motion Analysis) are used to derive the lower limb joint moments to be used as correct values. In addition, wireless wearable inertial sensors (MTw2: Movella) consisting of 3-axis accelerometer, 3-axis gyro sensor, and 3-axis magnetometer are used to acquire acceleration data to be used as training and validation data and test data. Ultimately, only one inertial sensor common to all three joints is used during estimation, and only that one axis is used. Therefore, in order to determine the suitable inertial sensor mounting position, the inertial sensors are mounted at four locations (Figure 2): pelvis, thigh, lower leg, and dorsal foot during the data acquisition experiment.

The experimental subject walks on the walking path; 50 trials of 10 steps per trial are performed. Three force plates are placed on the 5th to 7th step (steady walking) of the walking path, and the subject is required to take only one step on each of these force plates. The sampling frequency of each device is uniform at 100 Hz.

After the walking experiment, lower limb joint moments are derived using inverse dynamics analysis software (KinTools RT: Motion Analysis) based on a total of 29 three-dimensional coordinate positions on the whole body obtained from the optical motion capture system and three-dimensional ground reaction forces from the force plates. The obtained lower limb joint moments are values in the world coordinate system. On the other hand, the acceleration output from the wearable inertial sensor is data in a local coordinate system that has been corrected for the motion caused by wearing the sensor. Therefore, using the angular velocity data obtained from the inertial sensor, the acceleration was also prepared as data converted from the local coordinate system to the world coordinate system. When using acceleration in the world coordinate system, high estimation accuracy can be expected because the coordinate system is unified with the joint moments. On the other hand, when using acceleration in the local coordinate system, estimation can be realized with acceleration sensors alone, without using inertial sensors, because angular velocity data used only for coordinate conversion is unnecessary. In general, acceleration sensors are less expensive and easier to obtain than inertial sensors. We believe that this is an advantage.

All data obtained were smoothed by low-pass processing with a cutoff frequency of 9 Hz.

In the present study, only the stance phase, which causes ground reaction forces and places a high burden on the joints, is considered in the range of estimation. The left leg is the target. Vertical ground reaction force data obtained from the force plate were used to determine the ground and release times during the stance phase of the left leg. Based on these times, joint moments and acceleration data for the stance phase of the left leg only were extracted and combined, respectively. Since the number of data differs from trial to trial, the acceleration data and the respective lower limb joint moments for approximately 40 of the 50 trials are used as training data, and those from the 41st trial are used as validation data. For the remaining nine trials, the acceleration data is used as test data and the lower limb joint moments are used as correct values for accuracy verification. As an example, Figure 3 shows the acceleration in the walking direction in the world coordinate system at the dorsal foot of subject A and the hip joint moment of the left leg. The green dashed box is the training data.

The learning algorithm for deep learning is Long Short-Term Memory (LSTM), which is suitable for time series waveform estimation. This decision is the result of comparing LSTM, Recurrent Neural Network (RNN), and Gated Recurrent Unit (GRU) in prior experiments. The hyperparameters determined by trial and error are shown in

Table I. The number of input data was set for each subject based on the estimated correlation coefficients for nine trials. The appropriate values for subjects A and B were 65 and 62, respectively.

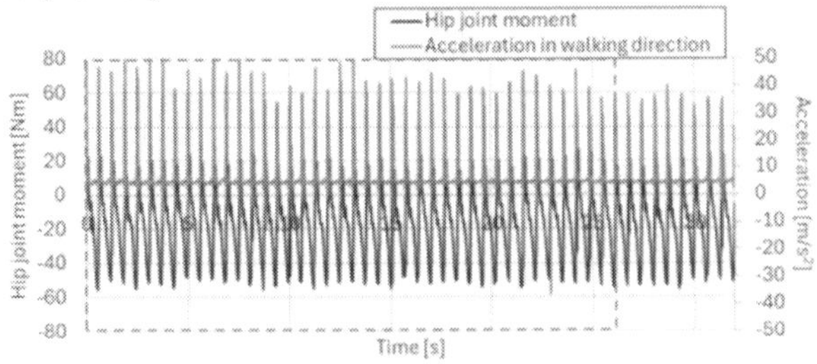


Figure 3. 50 trials of single-axis acceleration measurement data and joint moment calculation data.

TABLE I. LEARNING CONDITIONS.

| | |
|---|---|
| *Number of hidden layers* | 50 |
| *Number of epochs* | 50 |
| *Batch size* | 32 |
| *Learning rate* | 0.001 |

## III. RESULTS

To determine the mounting position of the inertial sensor that obtains single-axis acceleration data, we performed individual learning for subject A with all 3-axis acceleration data in the world coordinate system obtained from 4 inertial sensors for each of the three joints. Then, we estimated with unknown test data for subject A. As results, the dorsal foot acceleration in the walking direction was selected from the twelve data points. This is because a balanced and high estimation accuracy was obtained for all three lower limb joint moments in subject A. In this section, the results are presented.

The trained deep learning models for each subject were created using training and validation data, which were acceleration in the walking direction obtained from the wearable inertial sensor attached to the dorsal foot. Subsequently, the joint moments of the left leg were estimated three times for nine trials using each test data for each subject. Table II shows the correlation coefficients and mean absolute errors with the joint moments calculated

TABLE II. ESTIMATION RESULTS.

| *Subject* | *CS*[a] | *Joint moment* | *Correlation coefficient* | *MAE*[b] *[Nm]* |
|---|---|---|---|---|
| *A* | *World* | *Hip* | 0.948±0.0066 | 4.47±0.375 |
| | | *Knee* | 0.972±0.0020 | 3.53±0.253 |
| | | *Ankle* | **0.985±0.0055** | **3.82±0.665** |
| | *Local* | *Hip* | 0.946±0.0035 | 4.53±0.313 |
| | | *Knee* | 0.969±0.0038 | 3.97±0.541 |
| | | *Ankle* | **0.987±0.0044** | **3.85±0.505** |
| *B* | *World* | *Hip* | **0.943±0.0006** | **6.88±0.205** |
| | | *Knee* | 0.948±0.0032 | 4.76±0.296 |
| | | *Ankle* | 0.975±0.0064 | 7.48±0.821 |
| | *Local* | *Hip* | **0.938±0.0090** | **7.70±0.798** |
| | | *Knee* | 0.939±0.0084 | 5.23±0.538 |
| | | *Ankle* | 0.975±0.0050 | 9.44±1.225 |

a. Coordinate System b. Mean Absolute Error

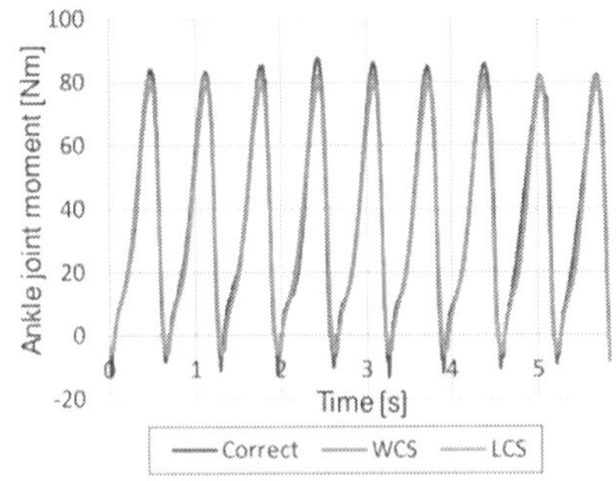


Figure 4. Estimated and measured ankle joint moments for subject A in nine trials.

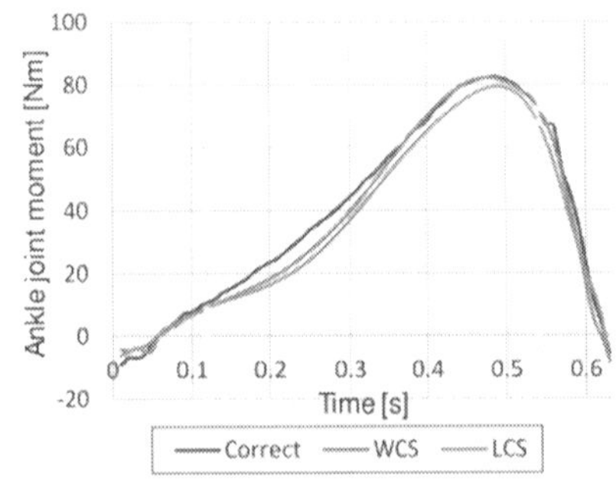


Figure 5. Estimated and measured ankle joint moments for subject A in one trial.

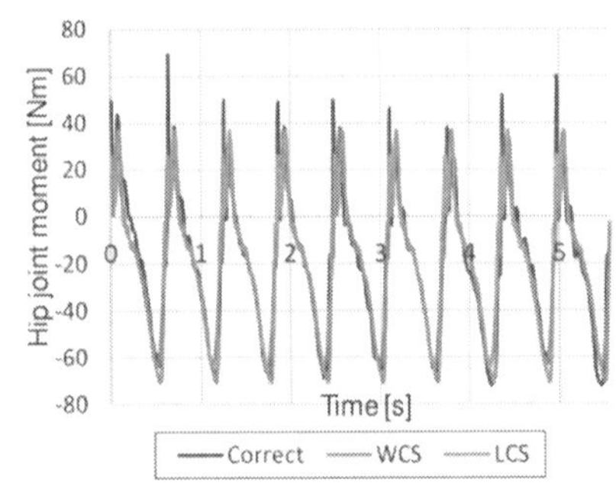


Figure 6. Estimated and measured hip joint moments for subject B in nine trials.

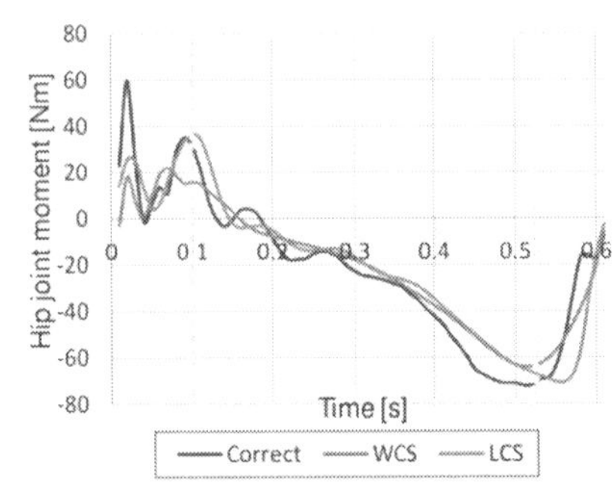


Figure 7. Estimated and measured hip joint moments for subject B in one trial.

based on actual measurements as the correct values. Figure 4 shows the results of nine trials of ankle joint moments for subject A, and Figure 5 shows only the ninth trial of Figure 4. From Table II, this was generally the highest correlation coefficient among all the estimations. Figure 6 shows the results of nine trials of hip moments for subject B, and Figure 7 shows only the ninth trial of Figure 6. From Table II, this was generally the lowest correlation coefficient among all the estimations. In these figures, the blue line shows the estimated values using acceleration in the world coordinate system, the gray line shows the estimated values using acceleration in the local coordinate system, and the orange line shows the correct values. The stance phase begins with the double support phase, passes through the single support phase in which the other leg (the right leg in this case) is in the free leg phase, and ends with the double support phase in which the other leg is grounded again. In Figures 5 and 7, the yellow dashed box indicates the single support phase, and the others indicate the double support phase.

In Table II, the correlation coefficients between the correct and estimated values are all above 0.9, indicating the presence of a relatively strong positive correlation. Furthermore, the strength of the correlation can be observed in Figures 4 and 6. The mean value of MAE presented in Table II is 7.4% of the mean body mass, which is small, and the standard deviation is 0.74%, which is also small. In other words, Table II and Figures 4 and 6 demonstrate that the results for nine trials were highly accurate. Figures 5 and 7, which show one trial, indicate the result for the single support phase is generally consistent, but there are steady-state errors and errors that do not follow minor changes in the double support phase. As the double support phase in one gait cycle is short and the ankle joint moments, as shown in Figure 5, vary gently, so errors in the double support phase are not a problem. However, for the hip joint moments, as shown in Figure 7, the failure to capture the peak values in the initial double support phase may have implications. This is because, as previously stated, joint moments can be used to represent muscle activity, with the peak value representing the maximum load on the joint. Therefore, two sources of error and suggestions for improvement are listed below. The first is that most of the stance phase is during the single support phase, and there are no large moment fluctuations during this phase at any joint, so the number of input data determined from the overall correlation coefficient was biased toward the larger values. Second, because only the stance phase was extracted and combined, there were discontinuities at the trial junctions. We believe that by setting the estimation range to one gait cycle that includes not only the stance phase but also the swing phase, in which the moment is zero, continuity will be maintained and errors will be reduced.

Besides, from Table II, both the correlation coefficient and MAE are slightly less accurate for subject B than for subject A. As mentioned earlier, this is due to the fact that the hyperparameters were set and the sensor mounting positions were determined using data from subject A. In addition, early stopping was not used in the present study. Therefore, there is a possibility of overfitting in the learning of both subjects, especially in subject B. Optimization of hyperparameters and sensor position, in addition to incorporation of early stopping into individual learning for subject B would have yielded better results. However, the results for subject B also showed good values, which means that even if the parameters were optimized for other subjects to save time and effort, good results could be obtained with a healthy gait.

In addition, comparing the results in the world coordinate system with those in the local coordinate system, there is no significant difference. Therefore, this study adopts estimation using a local coordinate system, which requires only one sensor for measurement and no coordinate transformation during estimation.

## IV. Conclusion

This study examines a convenient method for estimating quantitative parameters useful for self-healthcare. Therefore, in this paper, the three lower limb joint moments were considered as effective parameters, and a convenient method was proposed to estimate them using a trained LSTM model by measuring only the actual single-axis acceleration data. As its acceleration data, we decided to use the dorsal foot acceleration in the walking direction, which provided high estimation results for all three joint moments simultaneously. From the estimation results of individual learning for each of the two subjects, although some errors remained during the double support phase, the overall estimation in each of the two subjects was highly accurate, regardless of whether a world or local coordinate system was used for the acceleration data. Thus, it is expected to be possible to verify the effect of exercise by simply installing a small and lightweight acceleration sensor during daily walking exercise, without restrictions on time and place.

In the future, the generalization performance will be evaluated with an increased number of subjects in order to improve the practical relevance of this study. Furthermore, we will apply the proposed method to other gaits.

## References


[1] S. Yamamoto, "Biomechanics of Human Movement," J of Physical Therapy Science, vol. 18, no. 3, pp. 109-114, Aug. 2003, doi.org/10.1589/rika.18.109. (in Japanese)

[2] S. Kawamura et al., "Inverse dynamics analysis of running motion by using wearable sensors," SHD2016, The Japan Society of Mechanical Engineers, vol. 82, no. 834, p. 15-00438, doi.org/10.1299/jsmeshd.2016.A-15. (in Japanese)

[3] H. Kotani, K. Shibata, Y. Inoue, and H Satoh, "Estimation of Hip Joint Moment by an Inertial Measurement Unit," 1st International Conference on Human Systems Engineering and Design (IHSED2018) Springer, vol. 876, pp. 946-951, Oct. 2018.

[4] H. Kotani, K. Shibata, and Y. Inoue, "Estimation of lower limb joint moment using few IMUs," The 58th General meeting and Conference, The Japan Society of Mechanical Engineers Chugoku-Shikoku Bracch, vol. 2020.58, p. 01b1, 2020, doi.org/10.1299/jsmecs.2020.58.01b1. (in Japanese)

[5] M. Mundt et al., "Prediction of lower limb joint angles and moments during gait using artificial neural networks," Medical & Biological Engineering & Computing, vol. 58, pp. 211–225, 2020, doi.org/10.1007/s11517-019-02061-3.

**International Academy, Research, and Industry Association (IARIA)**
412 Derby Way
Wilmington, DE 19810

ISBN 979-8-3313-1387-6

CURRAN ASSOCIATES INC.
proceedings.com

# Fourteenth International Conference on Autonomic and Autonomous Systems (ICAS 2018)

Held at InfoSys 2018

Nice, France
20 - 24 May 2018

IARIA